New
Sentinels
of the Seas

New Sentinels of the Seas

Satellite AIS and the Birth of Global Maritime Awareness

By Geo. Guy Thomas
CDR, USN (ret.)
Inventor, Innovator
and Developer

New Sentinels of the Sea:
Satellite AIS and the Birth of Global Maritime Awareness

Library of Congress
Cataloging-publication-in-data
ISBN 979-8-9853477-3-9
Printed in the United States of America

First Edition

Editors:
William Lambrecht
Sandra Olivetti Martin
New Bay Books • Fairhaven, Maryland
newbaybooks@gmail.com

Design by Suzanne Shelden
Shelden Studios • Prince Frederick, Maryland
sheldenstudios@comcast.net

Cover imagery:
Digital collages use the following two images:
Global Telecommunication Network Around the World
by NicoElNino, Shutterstock
Satellites and Networks, by Andrey VP, Shutterstock

Note on Type:
Titles and Headers set in Arial
Text is set in Garamond Premier Pro

Dedication

To my smart, beautiful, indefatigable wife, Clelia. It is her untiring support over the past 25 years that has given me the time and energy that has allowed me to pursue research on my concepts, and write both this book and my memoir of my time in the intelligence world, *A Silent Warrior Steps Out of the Shadows*, while she built, and now runs our historic, award-winning Wilson House Bed & Breakfast. I could not have done it without her.

"The best idea came from Guy Thomas. He said we should put AIS receivers in satellites and watch from space. This, he argued, would give us an almost worldwide view of the ships carrying AIS. After the meeting, I asked Guy how we could confirm that his idea was possible and how we might implement it (at the time I did not know Guy had been thinking and talking about this for years)"

—Jeffrey P. High
(Former director, US Coast Guard
Maritime Domain Awareness
Program Integration Office)

Foreword

Sentinels of the Sea: A New Solution to an Old Problem

Intelligence veteran George Guy Thomas, author of *A Silent Warrior Steps Out of the Shadows*, herein presents the inside story of his struggles in his post-9/11 crusade for security on the seas, which led to conceiving and implementing Satellite Automatic Identification System ship tracking and a concept he calls C-SIGMA (Collaboration in Space for International Global Maritime Awareness). Thomas, a quadruple qualified (surface, air, submarine, space) retired Navy commander, warns that American ports remain vulnerable as he presses for global cooperation in patrolling the seas from space for safety and security.

Thomas is nominated for the National Medal of Technology and Innovation, the nation's highest honor for technological achievement, "for overcoming severe obstacles and dramatically changing the maritime world by lifting the veil of opaqueness enshrouding the seas."

He was also nominated three times for the Space Foundation's Technology Hall of Fame, and in 2021, Thomas received their lifetime achievement award for his role in Protection and Increased Safety of the Maritime Environment. The citation asserts that as a result of his leadership, "commerce and supply chain operations can flow safely around the globe, while security threats and needed resilience to ever-dynamic risks are adequately addressed by our nation and its allies."

His efforts engendered increased global security, better protection of fisheries and the marine environment, and hastened recovery from disasters on or adjacent to the world's oceans. The innovations he brought about have been referred to as a paradigm shift in the maritime world on the order of the chronometer, the steam engine, the screw propeller, radar and satellite navigation.

Award presented to Cdr. Thomas

Praise for Thomas

"S-AIS is a fundamental technology that impacts every single American's life every day…It only exists because of Guy Thomas."
Keith Masback. *Former Chief Executive Officer of the United States Geospatial Intelligence Foundation*

"His extraordinary vision and persistence brought space-based Satellite Automatic Identification System (S-AIS) to the world. His writing has been a primary guidepost to not just us but the entire marine world."
Dirk Vande Ryse. *Director, Situational Awareness and Monitoring Division, European Border and Coast Guard Agency*

"Guy Thomas's creation and implementation of S-AIS and C-SIGMA critically support American and global maritime security, safety, economic progress, trillions of dollars in maritime trade, and environmental stewardship of the oceans."
Julio J. Gutiérrez. *Capt., US Navy (Ret.); Maritime Security Consultant*

Table of Contents

Introduction

The memoir of the first two-thirds of my professional life is titled *A Silent Warrior Steps Out of the Shadows*. I chose that title as the first 40 years of my professional life were in the "shadows" of the classified world. In the shadow world, I was what was known as a "silent warrior" because, in those days, you could not discuss your work with anyone, not even your family. I was in the US Navy for 23 years, for the last 20 directly involved in collecting intelligence as a member of the little-known Naval Security Group. The second 20-plus years of my professional life I spent in research and development directly related to intelligence collection, analysis, and dissemination to users in a "tactically useful timeframe."

I use the term "tactically" with care because I operated and worked at the tactical level as opposed to the strategic or operational levels most of my career. There are three levels of military operations. From the top down, they are "strategic," relating to the gaining of overall or long-term military advantage; "operational," relating to the overall functioning and activities of a military organization; and "tactical," relating to, or constituting actions carefully planned to gain an immediate specific military end. My career in the shadow world focused on the here and now, the tactical level, although the collection of intelligence spans all three.

My memoir took 19 months to receive publishing approval by the six organizations I had either worked for or with in my 40-plus year career in the classified world. Those six were the National Security Agency, the Central Intelligence Agency, the Federal Bureau of Investigation, the National Security Council, and the Navy, Air Force, and Coast Guard's intelligence branches. Views and opinions in this book are mine and do not reflect those of the NSA or other agencies. For the past 15 years, I have consciously avoided the classified world and concentrated on unclassified space systems. How and why I made that transition from the shadow world to the unclassified commercial space realm is an excellent place to start this story. But let's set the stage.

A well-worn adage advises, *Find a job you love, and you will never work a day in your life.* That is precisely what happened to me. I left a prestigious, well-paying job as a researcher at Johns Hopkins' Applied Physics Lab to take a tax-free $25,000 a year cut in contributions to my retirement account to go to work as a civil servant. I do not regret my transition from the classified to unclassified world in any way.

I originally planned to include this story as part of my memoir, but I decided to publish this part of my life as a separate book based on many people's advice. This allowed me to provide a focused history of the start of Maritime Domain Awareness, Satellite Automatic Identification System (S-AIS), and space-based Maritime Situational Awareness, endeavors that are still very much in play on the world stage today.

In separating the two books, I also hope to help foster focused discussion on the increasing utility of Earth observation space systems to dramatically improve safety and security in the maritime domain and increase protection of the marine environment and its resources. Global Maritime Awareness, as I call these efforts, will also aid in marine-related disaster mitigation and recovery and significantly help improve the marine transportation system.

From 12 September 2001 (the day after 9/11) until 22 April 2005 (the first C-SIGMA meeting), I concentrated on the many maritime safety and security issues related to counterterrorism, counter-piracy, and smuggling. Why will be discussed later. At the first C-SIGMA meeting, the Norwegian attendees introduced me to two other maritime world concerns: environmental and resource protection/conservation. Until then, I had not thought about the importance of environmental and resource protection. Indeed, I knew almost nothing about either of them. However, by the end of 2005, I knew these two issues to be huge, fundamental problems.

About the same time, I also realized one Global Maritime Awareness System could address all four concerns: security, safety, environmental protection, and resource conservation. It would be a boon to all involved in legal activities related to the maritime world and a significant hindrance to all involved in illegal activities at sea. It would also substantially increase marine safety in many instances.

Additionally, I recognized that S-AIS would provide geolocation and identification data that would be the crucial underpinning of this system. By then it was clear the efficient utilization of the rapidly expanding number and capabilities of Earth observation space systems, with S-AIS leading the way, was the key to developing a solution to all of these needs. Thus, since early 2006 I have focused on building one system to address all four maritime problems: Security, Safety, plus Environment and Resource protection. I have called this effort C-SIGMA: Collaboration in Space for International Global Maritime Awareness.

This book starts with a short story that illustrates why C-SIGMA is very important and then relates how satellite AIS came to be, continues on to how it has changed the maritime world forever, and concludes with a discussion of opportunities it offers us for the future. It is told from my viewpoint because C-SIGMA has been the primary professional focus of my life since I first had the idea to put an AIS receiver on a satellite in low Earth orbit and use it as the long-desired global ship tracking tool on 4 October 2001.

This is also the story of how Maritime Domain Awareness and space-based Maritime Situational Awareness came to be common concepts and capabilities in the maritime world today. The book ends with my best estimate of how these three capabilities—S-AIS, Maritime Domain Awareness and space-based Maritime Situational Awareness—will coalesce into Global Maritime Awareness (GMA). Then comes a final call to action to build the Global Maritime Awareness system via C-SIGMA.

This book is timely because our ports and coasts remain at risk. In the years just following 9/11, Homeland Security and the Defense Department were very concerned about unknown vessels entering our harbors undetected. Their concerns centered on illegal narcotics and aliens but also included saboteurs with weapons, up to and including, heaven forbid, nuclear and other weapons of mass destruction.

This last item, nuclear weapons, was not an unreal pipedream. We had reports that, at the dissolution of the Soviet Union in 1991, a number of small but powerful nuclear weapons had disappeared, no one knew where they were. Speculation was that they were in the control of Russian gangsters and might end up in the hands of anti-American terrorists. Probably the easiest, most effective way to get these weapons into the US, and place them where they could do the most damage would be to put them on small ships, possibly fishing boats, and sail them into one or more of our harbors.

The terror weapon did not need to be a nuclear one. Chemical and biological weapons could also be highly effective in shutting down a major harbor such as San Diego, Houston, or even Los Angeles/Long Beach.

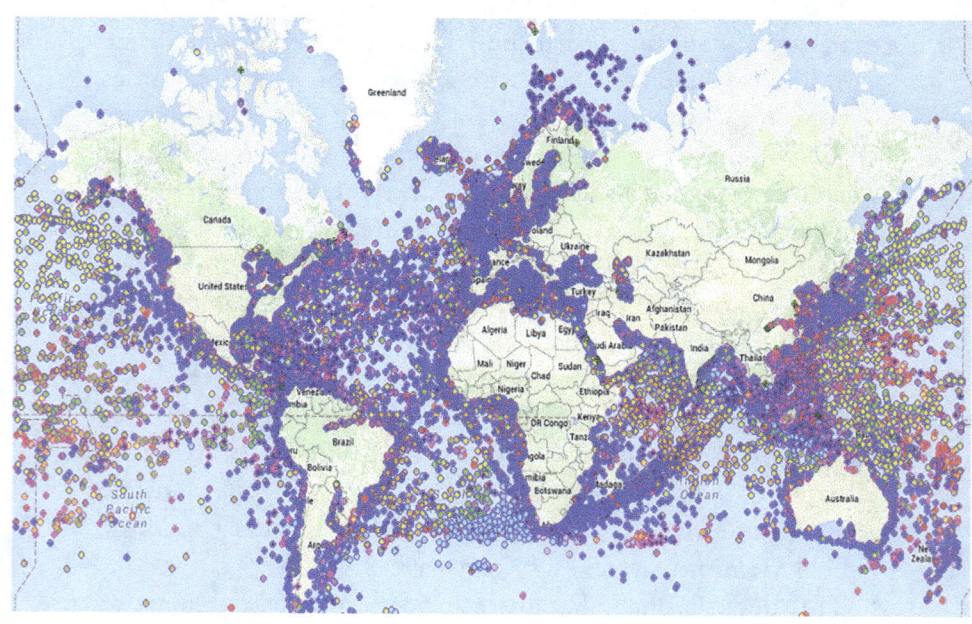

A day's collection with the ship type shown by colors.
Courtesy of ORBCOMM.

Our list of potential targets was the 18 major ports and a couple of hundred smaller ones. After James Woolsey, a former director of the CIA, gave a speech in San Diego over 15 years ago citing this as a major threat, he and I had a short discussion on this subject. He pointed out that while he was with the CIA, we had lost track of two tractor trailer loads of anthrax in Iraq, just before the war broke out. He suspected they were smuggled into Syria, which was much more stable in those days. He further pointed out that a half-gallon of anthrax exploded on a boat in San Diego harbor could lock down the city for months.

From 2005 to 2007, I participated in several studies focused on how we could detect, track and identify all vessels, from mammoth ships transporting oil and cargo, down to vessels of 20 to 25 feet. We were especially concerned about the smaller vessels. Indeed, the last study I participated in for Homeland Security Science & Technology focused on the small boat problem lasted from 2015 to 2019. It was canceled only when its funds were diverted to build a wall on our southern border.

Why we have not built the system to counter this problem baffles me. A global system to address the problems of the maritime domain needs to be built. Given the demonstrated animosity of some groups, it is especially needed in the United States.

Chapter One

An Attack

The vessel was unremarkable, a working boat of some sort, with ample evidence of having spent much more time at sea than in maintenance yards. Indeed, it had a worn look about it as it moved past the sea buoy marking the entrance to the channel into San Diego Bay and turned east toward the one of most beautiful cities in the United States.

San Diego is also the home for a major part of the US Navy's Pacific Fleet, and that was one of the reasons this nondescript vessel had been tasked by decision-makers half a world away to enter the picturesque harbor before dawn this day. That was not the only reason to select San Diego. It is also a major commercial harbor, as are the five other ports along the Pacific Coast where the same basic scenario was being enacted that morning. In Long Beach, Oakland, Seattle, Vancouver and Prince Rupert, equally unremarkable vessels of medium size, each less than 100 feet and 300 tons but clearly ocean-transit capable, entered its assigned port at basically the same time.

The working boat was about 75 feet in length and 100 tons displacement. The size had been carefully selected. The superb seamen of Southeast Asia who plotted and were executing this operation were very aware that the International Maritime Organization requires all ships over 300 tons to comply with many ship tracking and safety regulations. All maritime nations of the world abide by these regulations. With vessels that did not come under those regulations, the chances of detection by the Navy or Coast Guard of the United States or Canada were significantly reduced.

The vessels were pirated or purchased, one by one, in the South China Sea or the approaches to the Straits of Malacca, hidden in the rivers and estuaries of Indonesia and the Philippines, carefully checked to be sure they could make the one-way trip to the West Coast of North America and placed in readiness. Finding crew was not a problem, for Indonesia is home of the largest group of radical Islamists, with a large number in the southern Philippines as well. All were excellent seamen, hardened to privation, having spent many years at sea on fishing or small cargo boats on which most westerns would not set foot.

The six boats had gathered in the eastern approaches to the Celebes Sea and crossed the Pacific Ocean, the largest body of water in the world, in very loose formation, miles apart. Each vessel was equipped with the normal radar associated with its type, AIS, GPS and two satellite phones. The GPS was in constant use to maintain track and relative position. The AIS was used only in a receive mode. When it warned that an unknown ship was approaching, radar was turned on to keep track of it and subtly adjusted their course to avoid all contact if possible. The satellite phones were used hardly at all, and then only with very short transmissions. The same phone was never used twice in a row. The six vessels, traveling well outside any shipping lanes or fishing grounds, had taken pains to appear as if they were not sailing together. They had rendezvoused only twice, once north of Eniwetok and Bikini, and once well north of Hawaii. One of the vessels was a tug, and it towed a barge loaded with enough fuel and food for the six vessels to reach their targets with ease. After the last rendezvous and all remaining fuel and food were off-loaded, the barge was scuttled with charges placed in its hold. It sank quickly, and the six vessels dispersed to meet four days later 275 miles off northern California for one last time. At this meeting, final plans were reviewed and all were checked to ensure they were ready in every way to carry out their "divine" mission.

There was some minor difference in arrival times at each port, and that was no accident. Each arrival time had been carefully calculated. Each had a specific target, and the time needed to travel the distances from the harbor entrance to their assigned point was different for each, thus requiring different entrance times. Each vessel needed to enter its assigned harbor in the predawn darkness because they did not want to be observed too closely as they each dumped what appeared, to any untrained observers, to be a series of 55-gallon drums over the side in the center of each ship channel at several minute intervals. Some of the dumping was delayed as the crews took pains to not be observed from any passing ship. That did not bother the captains involved as randomness was a virtue in this case.

The devices were mines, with timers set to arm them two hours after they were laid. However, mines were not, even as deadly as they were, the most deadly cargo these vessels carried. On each ship's deck was a loose sack, made of very heavy gauge rubber and hermetically sealed, with between 25 and 50 pounds of anthrax, along with an aerial dispersant. The sack rested on a four-foot square steel plate with a wire mesh rim. The whole vessel was rigged as a bomb, and the intent was for the plate to be blown into the air, for the mesh to rip the bag open and for natural forces to do the rest, literally scattering the deadly cargo to the winds.

At exactly 0740, each of the ships arrived off its assigned spot which, with one exception, was calculated to be as close as possible to the major cargo-handling facility in each city. That one exception was San Diego, where the nondescript boat positioned itself upwind from the aircraft carriers tied up at North Island and then, turning sharply, made straight for the nearest carrier. The Navy small craft assigned to warn off civilian boats that got too close to the carriers immediately noted the change in behavior of the unremarkable ship and, increasing speed to its maximum, moved to intercept, broadcasting a warning both on loudhailer and on the harbor common radio channel ordering the unknown vessel to change its course or be fired on. But the warning was too little, too late. The ship's first defense, a floating barrier stopped the vessel in its tracks. Thirty seconds later the Navy guard boat pulled alongside, its crew with M-4s locked and loaded, ready to board the vessel piled on the floating barrier. The crew never had the chance.

The explosion was heard all the way to Tijuana to the south and Carlsbad to the north. The Navy boat had done its mission and the carrier was untouched, but the cloud generated by the explosion was far more deadly than anyone imagined. Sailors and officers alike clamored to the side of nearby Navy ships to gawk at the smoking ruins. Likewise, seamen on the commercial fishing boats and merchant ships and thousands of people all around the bay stopped to stare at the column of smoke rising from the ruined ship lodged on the carrier's floating barrier. No one had any idea that a week later many of them would be dead, succumbing in a most horrible manner, or that the economy of the United States, and thus the world, would be devastated.

Each of the other five vessels also carried out its mission, and anthrax was now widely scattered over the six major West Coast ports. Additionally, in each port at least one of the newly laid mines had performed its task and badly damaged or sunk a merchant vessel. In San Diego, the first vessel to hit one of the mines was a visiting Japanese Navy Helicopter Destroyer. Its crew was preparing for the continuation of its goodwill visit to Canada, the United States and Mexico that morning and had quickly gotten underway at the sound of the explosion. The captain had placed the ship at general quarters as a precaution and thus it was saved from sinking. But 26 of the crew died and the ship was barely able to return to a berth at the Navy base. It would be many months before it saw the circling ospreys that welcomed ships entering Yokosuka, its homeport near the mouth of Tokyo Bay.

Even though the death toll of the explosion and then the subsequent anthrax poisonings was limited to a few thousand, the panic engendered by the attacks caused all six cities to halt most services. Chaos ensued, both in the ports and in the world's stock markets. It would take several years to recover, and when it did the world as we now know was completely changed. The economies of the United States and Canada were badly damaged; Europe, Russia, China and Japan also suffered substantial losses. India, Brazil and South Africa exhibited more clout on the international stage.

Al-Qaeda, praising a new ally in the radical Islamists of Indonesia, claimed responsibility for this most devastating of all attacks on the non-believers of the world. The West Coast attack, like 9/11, had taken months of planning. But it is the belief of the faithful that time is on their side. When it comes to maritime vulnerability, they could be right.

Chapter Two

When the World Works Together

The West Coast attack did not happen, of course, due in large measure to the subject of this book— Satellite Automatic Identification System. Maritime Domain Awareness and Satellite AIS has been my passion for two decades, and I believe the fruits of my labor would be the reason for an alternative fate that follows for those ill-intentioned vessels gathering in the Celebes Sea.

A Finnish SAR (synthetic aperture radar) satellite on a routine collection over the Papahānaumokuākea Marine National Monument, the vast expanse of water west of the Hawaiian Islands extending over 1,000 nautical miles past Midway Island, detected three of the vessels on a course that could only take them to the Aleutians Island Chain in Alaska. An anomaly detection software tool called PANDA, originally developed by the Defense Advanced Research Projects Agency, noted the unusual course of the vessels. An alert analyst at the watch center in Japan thought it odd that three ships would be bound for that remote part of the Pacific at the same time yet not traveling together.

The analyst's suspicion triggered an initial investigation using space-based AIS collected both by United States and Canadian commercial satellites, then the output from the newly launched commercial electronic intelligence (ELINT) satellites from the United States, England and France, plus reports from the Maritime Safety and Security System. Fusion and analysis software tools initially developed by the Defense Department's Joint Capability Technology Demonstration (JCTD) program and several similar programs in Europe, first under the European Commission's

LIME land and sea integrated monitoring for European security assessment framework and then MARISA—Maritime Integrated Surveillance Awareness—enabled a rapid search and correlation of data by artificial intelligence and machine learning. It quickly became obvious that the ships were not behaving in a normal manner. The vessels were not broadcasting AIS, and they appeared to be avoiding the Long Range Identification and Tracking (LRIT) reporting requirement area. Thus they did not have to relay their position every six hours, further heightening interest.

More analysis of open source information available at the watcher's operations center quickly established that the ships were "unknowns." The operations center evolved from a series of bilateral and multilateral agreements, part of a new era of international collaboration being tested on this day. The watch supervisor forwarded a request for additional collection from the next available synthetic aperture radar satellite, which happened to be Italian, as well as optical imagery and ELINT collection. Intelligence organizations of the northern Pacific were alerted and swung into action, each in its own way. The tasked commercial optical systems included one from the United States, one French, and one Japanese. SAR images revealed a total of six vessels on this unusual track in a loose trail formation, and the optical systems determined them to be nondescript vessels. Still, these ships did not seem to be on a threat vector to anywhere, so it was decided to watch them as they proceeded northeastward across the Pacific.

Two days later, when the ships were missed on a scheduled collection from an Italian 5 SAR satellite constellation, all available space assets were immediately tasked to relocate the vessels. This included 18 nations, thanks to C-SIGMA. The correlated collection effort paid dividends; the suspect vessels were reacquired less than 12 hours later. It was obvious now that the ships had turned east and started closing the North American coast, near the US-Canadian border. To say interest increased greatly is an understatement.

Still, the authorities had nothing specific to go on other than rumors of a terrorist plot involving stolen vessels coming out of Southeast Asia. They could be drug, arms or people smugglers, or perhaps terrorists. It was decided to intercept and investigate the vessels with a US submarine that happened to be returning from a northwestern Pacific deployment, and to deploy both US and Canadian naval forces (Navy and Coast Guard) to be in position to intercept, should that become necessary. Even though the operation had huge national security implications, in the final analysis this was really a law enforcement activity. A US Coast Guard one-star admiral who headed the service's special forces was designated the Officer in Tactical Command. His command ship was the Navy's USS *San Antonio* (LPD-17), with a very modern, high-capacity wideband communications suite. So the Navy was somewhat mollified. Both the Air Force and the Army had to agree, even though each believed they could handle the job better.

The submarine, using its sonar to acoustically detect the vessels, closed in, taking periscope photos and relaying them via satellite communications to the Office of Naval Intelligence, which in turn was working in concert with its international partners. The process quickly identified several of the vessels as having been reported stolen in southeast Asian waters. The decision was made to intercept and board before they could get within 150 miles of the North American coast. The question was how. Shortly after the decision to act was made, the submarine reported the ships were rendezvousing and refueling. Then the submarine observed and reported the transfer of unknown objects between the several ships and the supply barge being scuttled. As the suspect vessels departed on divergent courses, the submarine established and maintained an acoustic track on each vessel long enough to ascertain its course and speed. It immediately became clear each vessel was headed to a different major West Coast port of the United States and Canada.

Early on, it had been decided not to shadow the vessels with aircraft, as that might tip them off that they had been detected. The major worry was whether the vessels had weapons of mass destruction onboard. But there were other concerns as well. Indeed, the ships might just be carrying the largest haul of heroin and opium ever seen. Or, if they were terrorists, the value of taking them alive and being able to interrogate them and exploit the information found onboard, potentially could be huge.

They had to be stopped and boarded in such a way as to not harm any US or Canadian servicemen. But clearly, they could not be allowed to get close enough to shore to blow themselves up and—if they were armed with weapons of mass effect—assault the mainland.

It was decided to board the vessels at night, eight hours before they would arrive off each port. US Navy SEALS, Canadian Special Boat Forces, and US Coast Guard Deployable Operations Group (DOG) forces were dispatched via as many special mission helicopters as were immediately available to the five US Navy, three US Coast Guard and two Canadian Navy ships tasked with interdicting the vessels, now strongly suspected of being terrorist-manned.

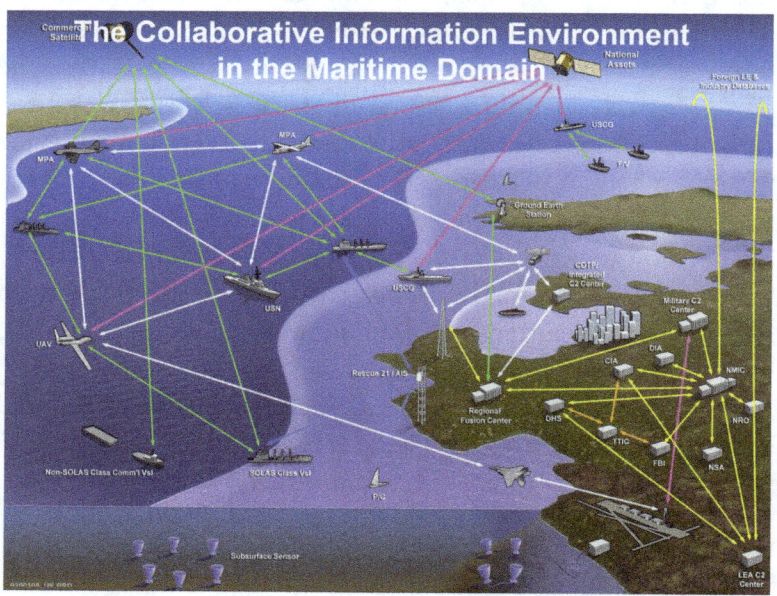

The original Maritime Domain Awareness diagram of 2001 as modified in early 2003. Artwork by JHU/APL. Courtesy of the author

Four of the helicopters were semi-stealth special mission capable, and six other regular mission helicopters were pressed into special service and positioned on the remaining task force ships, which were scattered from off Vancouver British Columbia to the Channel Islands off Los Angeles. The operation was planned for 0245 local, in darkness to maximize surprise and take advantage of the excellent night vision technology enjoyed by the special forces. A complete special force assault team was assigned to each target vessel. It was a good thing 10 helicopters were assigned because two had to abort due to mechanical problems. But the assault teams boarded without significant problems.

Total surprise was achieved, and only very limited resistance was encountered except when one man engaged the DOG team and was quickly neutralized. No friendlies were killed or even wounded. The mines, explosives, and the anthrax that could have proven so devastating were disarmed and rendered safe. The vessels, with the terrorists locked below, were escorted into port where they, their computers and other records provided a trove of information to the intelligence analysts. Japan, Canada, and the US congratulated one another, and the world celebrated the success.

Even the Russians published some very interesting periscope photos of the terrorist vessels taken from one of their nuclear fast attack submarines that, "just happened to be in the area," off the West Coast of the US. They'd been ready to assist if needed, they said.

Chapter Three

A Mission Takes Shape

In the hours after 9/11, the urgent imperative of Maritime Domain Awareness became clear.

As I walked into my office at the Naval War College on 9/11, I was greeted with the news that an airplane had just hit the World Trade Center. My office was a classroom that had been converted into five cubicles with a TV in the center of the room. Several people were gathered around it watching the news.

I had just returned from being the technical leader for Fleet Battle Experiment India, a live, at-sea test of 54 leading-edge technologies. The planning, execution, and first-look analysis had taken almost 18 months and countless road trips, so I was not embarrassed to be a few minutes late to work that day.

As we listened to the news and speculated why a plane would hit the World Trade Center, we watched in near disbelief as the second plane hit. I had been thinking the first crash was probably due to the pilot blacking out. With the second hit, I knew it was terrorism, and I immediately thought *We are at war! But with whom?* I knew then that life for everyone in the building was going to change significantly, but I had no idea of the extent of change in store for me.

My direct involvement with what is now called Maritime Domain Awareness began the very next day, in response to a direct order from President George H.W. Bush. On the afternoon of 9/11, he ordered the Chief of Naval Operations (CNO) to conduct a vulnerability analysis of the maritime assets of the US to terrorism. He also wanted a plan on ways to mitigate our vulnerabilities. The order came down the chain of command like a lightning bolt and, less than 24 hours later, I became the de facto technical director of the United States effort to understand our vulnerabilities to maritime terrorism and develop ways to protect ourselves. From that day until I went to work as the Science & Technology Advisor for the Coast Guard-led, interagency Maritime Domain Awareness Program Integration Office (MDA/PIO), I had three tasks. They were, in priority order: First, research, write, and work to implement a concept of operations to counter maritime terrorism aimed at the United States. Second, establish what became the Maritime Domain Awareness Program. And lastly, create what became known as Satellite-AIS.

I envisioned the Maritime Domain Awareness Program as a joint program office, not unlike the JPOs for the Joint Strike Fighter or the Joint Tactical Data Link, both organizations I had worked with very closely in past assignments. The idea for the MDA JPO came out of the wargames and seminars we held at the Naval War College in the several months after 9/11. I had a role in assembling and organizing those meetings to answer President Bush's directive to understand the vulnerabilities of the US to maritime terrorism and develop ways to mitigate them.

Vulnerability analysis and mitigation development are closely related but fundamentally different. The first part was clearly enough defined within the first three months, but the second part of the task, mitigation, consumed the rest of my professional career. There are two parts to mitigation development: Technology, both hardware and software; and the mix of policy, organization and administration called governance. In any challenge of this magnitude, it is almost always the last part, the governance portion of the problem, that proves the hardest to solve.

S-AIS was hardware, and the envisioned Maritime Domain Awareness Joint Program Office was an organization, part of governance. I believe I initially succeeded at both. But from late 2007 on, the program was hampered by politics, and two years after that, with poor, self-serving leadership, it failed. I will describe later in this book what transpired. Still, it was a noble effort, and I am proud to have been part of it.

I certainly did not create either S-AIS or the MDA/PIO by myself. However, the creation and advancement of both were my primary focus for the last 11 years of my professional life. The combination of the two into the C-SIGMA concept evolved from April 2005 to early 2006. I gave my first talk on C-SIGMA in March 2006.

C-SIGMA continues to be the leading cause for why I have not retired to my comfortable condo on the Pacific Ocean in Peru. As I stated earlier, C-SIGMA, short for Collaboration in Space for International Global Maritime Awareness, is a direct outgrowth of the national effort. But it did not come into play until four-plus years after Marine Domain Awareness became an established global concept. C-SIGMA will be discussed in more detail in its own chapter because Maritime Domain Awareness deserves full attention on its own.

I did not start as the head of the Naval War College's effort. That was Dr. Dan Smyth. He was the Center of Naval Analysis representative to the college, and we shared an office. However, his assignment to the college had already reached the absolute maximum number of possible extensions, and regulations forced him to return to CNA at the end of 2001. The report to the president was still very much in draft, and since I had been leading the writing effort since mid-October, it was natural that I assume the leadership role.

As the man charged with writing the report, I remained deeply involved. I very much wanted to, due in part to the need to continue to develop and implement the concept of operations beyond just the report and to follow-up on the ideas that became Long Range Identification and Tracking and S-AIS.

My tour was scheduled to end in January 2002, but I asked for and received permission to extend in Newport for one year, until January 2003. I later received approval for yet another extension to January 2004. However, that extension was cut short when, in May 2003, I was asked to join the newly formed Maritime Domain Awareness Program Integration Office (MDA/PIO) in Washington as soon as possible.

A tale of deception

In 1998, the US Coast Guard wrote a concept paper entitled Maritime Domain Awareness. That term has become recognized globally in the maritime world, and I helped in propagating and promoting it. Starting in the spring of 2002, I began collaborating closely with the US Coast Guard headquarters via Commander Gordon Thomas, the USCG lead for Maritime Domain Awareness. Indeed, it was Gordon—phoning late one afternoon in March 2002—who introduced me to the term Maritime Domain Awareness. Until that time, I called our draft concept of operations the Maritime Traffic Tracking System.

Every time I made a significant change to that draft concept, I distributed a copy to all participants, over 1,100 people. The list started with all attendees at the several wargames and meetings we held immediately after 9/11 and grew from there. Recipients had distributed it further, often copying me on their distribution, and I had added those people to my master list.

I received comments on my several drafts from over 100 people all over the world. I tried to acknowledge all of them, often getting into detailed email exchanges about the concept's specific aspects. Those exchanges were often very enlightening. Someone had forwarded a draft of my Maritime Traffic Tracking System concept paper to Gordon, and he instantly realized it was remarkably similar in many ways to the Coast Guard's Maritime Domain Awareness concept from 1998.

He suggested we pool our efforts, and I immediately agreed. I liked the term Maritime Domain Awareness and asked if he would mind if I took it for our endeavors. He said he would be delighted, so I did, retitling our draft concept of operations and sending it out to my list of over 1,100 people in the maritime world. Thus, Maritime Domain Awareness took on a life of its own far beyond the Coast Guard.

Gordon invited me to accompany him to a scheduled visit to the US Customs' Air and Marine Interdiction Center on March Air Force Base, Riverside, California in early May. He wanted to examine their capability to track ships. That sounded perfect to me, and I immediately accepted. Until that moment, I thought we needed to build an unclassified section in the various Naval Intelligence watch centers, which I knew would not be easy. Using a law enforcement watch center, which has much fewer classification restrictions, seemed an excellent idea. Nearly 20 years before, I had spent over a year without a sensitive compartmented information (SCI) clearance, due to circumstances I will detail later in the book. During that time, I had worked with law enforcement in counter-narcotics border surveillance operations.

The bottom line from that visit to the Air & Marine Interdiction Center was that they could track nearly every aircraft approaching any border from northern Peru to the Arctic Circle. However, they only had minimal vessel monitoring capability. There was just no way to do so. No detection, identification, or tracking systems. No sensor systems at all.

We talked with the senior staff there, especially with Jim Durett, then a contractor, but the primary system developer for their aircraft tracking system, and Roger Caudill, their Special Projects Officer, about polling the INMARSAT communications systems for all ships over 300 tons as a means of tracking these ships. The International Maritime Organization (IMO) already required vessels of that size to carry this communications system, and it seemed like it would be easy to implement. Indeed, this had been the one requirement generated by the first wargame back on 24 September 2001. It eventually led directly to Long Range Identification and Tracking (LRIT) requirements. The four of us agreed it would be good to test how much information could be gleaned from ship tracks by knowledgeable trackers such as the Air & Marine Interdiction Center had in abundance.

They were willing, even eager, to undertake the test on a "not to interfere" basis and were very interested in adding this mission to their duties. From talks with many shipping companies, I knew they were willing to participate, but transmitting on INMARSAT costs money, and they wanted to be reimbursed. I designed the test and asked the Navy Warfare Development Command, where I was assigned, to fund the limited objective experiment. It would cost around $95,000, but we asked for, and initially got $110,000. To make a long story short, we eventually had the money withdrawn under the deception that the Command had run out of money for the year. My boss directed me not to discuss this with RADM Bob Sprigg, the admiral in charge, as Sprigg's mind was allegedly made up and there was nothing I could do.

I thought this was very odd at the time, and it turned out it was. I had briefed Admiral Sprigg on the project, and he had tasked me to go forward with the planning, so I knew he was interested. Still, I decided to do nothing about discussing it further with the admiral as my boss had been clear that I was not to do that.

However, about three months later, at Admiral Sprigg's farewell party, the admiral asked me how the ship tracking project was doing. I told him I was disappointed we had lost the money. His reaction was immediate. He told me he had a surplus of cash at the end of the year and had tried to find good uses for it. My boss had told him there were technical difficulties, and that was why the project was delayed. I assured him there were no technical difficulties and that he had been misinformed. That was the first of many times I experienced outright deception and dishonesty by people not seeing the utility of Maritime Domain Awareness to the Navy or the Coast Guard, much less the nation and, indeed, the whole global maritime world, if not the world in toto.

I remember one senior person there at the command telling me with some heat that "it was not the mission of the Navy to track all civilian ships approaching our coasts; it was to bomb caves in Afghanistan." Osama bin Laden was thought to be hiding in those caves at the time and going after him was exactly what the carrier battle groups were doing. However, the Navy is not a one-trick pony and could have easily done both.

A 'yahoo' on stage

The US Coast Guard, first under Commander Gordon Thomas, a recently recalled retiree, and then under CDR Rick Stanchi, set up a maritime security interagency working group with representatives from anyone in the US government who believed they had a stake in the maritime domain and wanted to participate. The Coast Guard invited Canada to join because they were already part of NORAD, which would be the home of the new

Northern Command, and Northern Command was going to be assigned a maritime warning mission. The area for this responsibility coincided with its traditional air warning mission, which included Canada.

Over the next year, I worked very closely with the Coast Guard officers. We decided to use my brief developed from the Presidential-ordered, counter-maritime terrorism concept paper as the core for their brief to the Commandant of the Coast Guard on their work. The exceptional graphics department at Johns Hopkins University's Applied Physics Laboratory had illustrated my brief. It looked very professional because it was. Those guys are the best I've seen anywhere. Still, my Coast Guard partners needed to change the brief a little and add a few tweaks.

Because this brief contained my proposal for the MDA Joint Program Office that I had been advocating from the start of this task, I was delighted to help. I used my APL budget to have the graphics department polish the Coast Guard brief. CDR Richard "Ric" Stanchi was the scheduled briefer, so he and I visited the graphics department. He requested several relatively minor changes, and the graphics artist inserted them on the fly. Ric was very impressed with the skill of the artist. In the end, we were proud of our joint product. It just sparkled with energy.

CDR Stanchi gave the brief to a roomful of Coast Guard admirals a few days later. The Commandant liked the brief so much that he asked Ric to present it again at the next joint Coast Guard-Navy leadership meeting, scheduled for a few weeks later, in April 2003. The meeting between the Commandant of the US Coast Guard and the Chief of Naval Operations went very well. The two leaders immediately decided to establish a joint interagency office, as recommended by the brief. (Yeah!) They also agreed to ask the White House's National Security Council to invite all other US government departments involved in maritime affairs in any way also to participate. In the end, the Departments of Justice, Commerce, Transportation, and State, all agreed to support the office. The Navy and the Coast Guard already represented the Departments of Defense and Homeland Security.

At that time, the Coast Guard shared its headquarters with the National War College. However, the college had just finished building a new classroom and office building, and they were in the process of moving out of the Coast Guard headquarters building. Thus, they had room and offered some of it up as office space for the new organization, whose initial official title was the Maritime Domain Awareness Program Integration Office, or MDA/PIO in military speak. It underwent several name changes in the next seven years before being disbanded in 2011. Its next-to-the-last title was the national Office of Global Maritime Situational Awareness. I always put the word *National* at the front of the title as it was indeed a national office. But the leadership rejected the idea of officially adding the word national to our title for reasons I never understood or even knew in any detail at all. Indeed, for its entire existence, many thought it was a Coast Guard office. That misunderstanding generated problems for its whole existence. The word *National* should have been added.

At their meeting, the Navy and Coast Guard heads agreed the Coast Guard would lead, with a Navy deputy, and the two top positions switching in the future. The service chiefs granted the Coast Guard the authority out of their joint funds to hire 45 people to help staff the new office. The Coast Guard immediately appointed Jeff High, an SES-3—a high civil service rank—to head it. Jeff High hired Mike Kichman, a retired Army Special Forces lieutenant colonel as his chief of staff, the job I was hoping to get. Mike had extensive law enforcement experience and had dealt with many senior Coast Guard officers in his military career. He also knew his way around the various agencies in the DC area very well. So he was a natural choice, even if it did sound a bit strange to have an Army SF officer in that job. However, given that one of the most significant recommendations for mitigating the maritime terrorism threat was to build a special operations force within the Coast Guard, the appointment did make sense. Jeff tasked Mike with staffing the office, and he got right on it.

About 10 days after that significant meeting, the Coast Guard held its annual technology expo in Baltimore. I went to several of the workshops but was not asked to speak. I also attended several different presentations on achieving marine domain awareness. I found them most interesting because they were nearly all copying my work, which I had been distributing widely for comments since soon after 9/11.

Indeed, in the last plenary session, attended by something over 500 people, a presenter from MITRE, a highly regarded defense-oriented think tank, put up my slide done by the APL graphics department, depicting the maritime domain with all the elements connected with different colored lightning bolts to indicate data flow from various sensors and platforms over numerous types of communications. Many presenters had used that graphic throughout the conference. Truly, I believe it was used in almost every one of the briefings that even mentioned marine domain awareness. However, I had kept my mouth shut. Imitation is the sincerest form of flattery, and by the end of that day, I was feeling very flattered.

I was quiet until, after complimenting the artwork, the MITRE presenter went on to say that no actual analysis had gone into the creation of the "pretty picture." I was sitting toward the back of the hall and took a quick nose count of people in the audience who had helped me with the research that created the so-called "pretty picture." There were over 50 people, roughly 10 percent of the people in the hall, who had assisted my work.

Many of them were easy to spot as they had turned around and quizzically looked at me during the presentation. We had spent almost a year doing just the type of research that the yahoo on stage was saying had never been done. We had written the paper but could get very few to read it—including the White House, who had originally asked for it.

I decided I needed to say something, so I raised my hand as soon as the time came for questions. The moderator, who I knew, immediately recognized me. I stood up, identified myself as the author of the "pretty picture." Then I went on to explain that there were over 50 people in the hall that had helped me do the in-depth analysis that the graphic represented. I also let him know that there was a 100-plus page paper at the Secret classification level, which outlined a concept of operations on how to use these systems and detailed precisely what each of the lightning bolt lines represents, down to its waveform, frequency, and bandwidth.

The concept paper also addressed how these systems on the graphic should be integrated to dramatically improve maritime awareness for the United States. I suggested he use our work for the basis of his, rather than "replough the same field." I offered to share the document with anyone with a Secret clearance and sat down.

I received smiles and "thumbs up" from all over the auditorium, but what happened next very much surprised me. As I rose to leave a few minutes later, a big guy that looked like an NFL tight end was coming toward me under a full head of steam.

"My name is Mike Kichman. I am the Chief of Staff to Mr. Jeff High, the new director of MDA/PIO, and he has just tasked me with staffing it," he said.

I replied I knew the name, as well I did. He went on to say several people had told him he should talk to me, and after my just completed remarks to the speaker, he fully understood why. He offered me the job as, basically, the chief technical officer of the new organization on the spot. I had just finished the job interview for the job I wanted in front of a very large audience. And that is how I came to work at Coast Guard headquarters for the Maritime Domain Awareness Program Integration Office.

Leaping into the fray

It took another six weeks of bureaucratic haggling to make it happen. I eventually resigned from the Johns Hopkins Applied Physics Laboratory because there was no way to reassign me there as their employee. That was too bad because their retirement package is very generous. My pay stayed the same, but I stood to lose $25,000 a year tax-free input to my IRA. That is a big difference compounded over nine years, but I would make that decision again in a heartbeat.

On the day I went for my exit interview at the Hopkins human resources office, the man conducting the exit interview asked why I had resigned.

"Because I am going to a fascinating job," I said.

"You misunderstand my question. Why didn't you retire?" he asked.

"Because I only have nine years here," I told him.

"We are a Navy University Affiliated Research Center," he said, "and thus, your 23 years in the Navy count toward retirement. You actually have 32 years here for retirement purposes. Here is a letter from you requesting retirement. I suggest you sign it, and we tear up your resignation letter."

"Where do I sign?" were my next words. Even as I was leaving, the Applied Physics Laboratory was looking out for me. Good organization! It was worth a good bit of money to me.

Thus, in mid-October 2003, two years into my presidential-directed task, I resigned from an excellent, well-paying technical research position at Johns Hopkins Applied Physics Laboratory, where I was immensely proud to work. I moved to take on the job to help guide what I had just had a significant hand in creating. While it did not pay as much as I had made, I do not regret doing so and am glad I did what I did, retirement bank account notwithstanding.

I liked my Johns Hopkins work very much, but the Maritime Domain Awareness Program Integration Office was my "child," even though I was not actually in the room at its birth. But I'd been there when it was conceived. Indeed, I was a principal "father," and most of the briefing slides were mine. I could say the concepts were mine, too. But they were the distillation of the work of over a thousand people, the attendees of the several wargames and symposiums held at the Naval War College in late 2001 and early 2002.

There is the old saying, "Find a job you love, and you will never work a day in your life." Such was the case here. I believed, maybe naively, that I had a near-unique set of skills and experiences that could make a significant difference, initially in the United States and its neighbors' security. However, as I became more and more involved, I realized that we were, in actuality, working to help secure the entire maritime world, including its environment and resources, for their rightful, lawful use for the benefit of all people on Earth. But I am getting well ahead of myself.

Chapter Four

Confronting Naysayers

The first thing we did at the new program office was requisition, with new boss Jeff High's assistance, the large, empty conference room next door to our office. We bought a set of movable eight-foot-high panels for the entire room. We divided the panels into Platforms, Sensors, Processing/Fusion Systems, Display and Decision Aids, and Dissemination/ Communications Systems.

We quickly filled the panels with the briefings with which we were being deluged. The boards organized our thinking (especially mine) and helped all of us focus on what was lacking. I called it the MDA War Room. Many of the briefs came right out of the Fleet Battle Experiments (FBEs) and DOD Advanced Concept Technology Demonstrations (ACTDs). Thus I had been working on, and with, their subject matter for the previous six years and was entirely up to speed on them before they ever came in the door.

Many times, High pronounced himself just amazed at my depth of knowledge. I kept pointing out to him that I had just spent nearly four years studying this exact technology: Two years with the Fleet Battle Experiments and then two years with the presidential counter-maritime terrorism task. It was more like six years if you count my last two years at Johns Hopkins working on pertinent ACTDs before I moved to Newport. And I had brought 36 years of experience in the maritime world, most of which involved sensors and the processing/analysis/dissemination of the information they generated, especially the last half of those 36 years.

I was not being modest. In my military career I had the good fortune to go from one interesting job to the next, with the previous job leaving me well prepared for each new, often unique, one-of-a-kind assignment. I had just happened into yet another set of circumstances where I was very well prepared for the job at hand. Indeed, many of the presentations given to the

Maritime Domain Awareness Program Integration Office were the same ones I had seen for the Fleet Battle Experiments, and our team of well experienced professionals there in Newport had analyzed them for their "Good, Bad and Ugly" components. The only difference was the title slides; literally, in many cases. So I was, thanks to being part of that team, very well prepared.

My first official trip outside of Washington was just before Thanksgiving, 2003, to Cambridge, Massachusetts, to the Volpe Center. Boston was going to be the site of the Democratic National Convention the next summer, which prompted a gathering at the Volpe Center of all the various governmental players in Boston's maritime area. There were at least two dozen organizations there that day, but I had an additional task. My boss wanted me to ask them to build a web-based software tool to gather all the AIS signals being transmitted by ships within line of sight (LOS), approximately 25 miles, during the convention and to allow them to be displayed in the command center. We also wanted to use that tool for the Republican National Convention to be held in New York City a month later.

The Volpe Center was adamant. They had developed a tool to guide ships through the Panama Canal and the St. Lawrence Seaway, they said, calling it the Vessel Identification and Positioning Systems (VIPS). They were sure it was a much better tool and wanted to use it rather than AIS. The first major problem, of course, was that they would have to build many more of their tool kits than they had in inventory. The box housing their hardware had to be placed on the bridge of the ship being tracked. This was clearly not practical. We had a polite but forceful discussion and agreed to disagree. They showed me the door. I did not bounce when I hit the street, but it felt like I did.

My first trip for my new boss, and I failed. I felt very low. But the story was not going to get any better by waiting, so I went right in to see the boss first thing Monday morning and told him of my failure to convince the Volpe Center to assist us. His first question was right to the point, a point that I had not covered because it had not been discussed with me before I left to go north.

"You did tell them that we were willing to pay them for their work, right?" High asked.

"No, sir! No one told me we had a budget for this task," I replied.

"We have $450,000 available, but I don't want to spend it all with them. I need to go up there in a week, and I will explain we are not asking for them to do this for free," he said.

We eventually provided them with both a handsome sum and technical assistance to build the device. It worked very well, so well it was adopted by the 6th Fleet about 18 months later and became known all over the world as the Maritime Safety and Security Information System (MSSIS). It won several prizes for innovation.

The comical thing about this is the Volpe Center's write-up put together to describe how MSSIS came to be. It said something to the effect that they did a diligent search and discovered a need for this tool. Then their team built it from scratch at a significant cost to the center. Total malarkey! To this day, some of the Volpe people evade me when we are at conferences as they know I know the real story behind this information system. The program manager is not one of them. He thinks it is a great story.

High-level meeting indeed

One of the first things Jeff High wanted to do was to hold a conference for all the US government organizations with a stake in the maritime domain. I suggested we use the Warfare Analysis Lab at Johns Hopkins University's Applied Physics Laboratory. I arranged the visit and APL put on its usual great show. High enthusiastically bought the idea and assigned me to work with the APL staff to stage the first and perhaps only government-wide marine domain awareness summit, set for May 7, 2004. It was an interesting six months and we all, both government and APL members, learned a great deal. One thing stands out to this day: Technology is the easy part when compared with politics and policy development.

The Warfare Analysis Lab has 100 seats. There are 12 seats at the head table and 88 in a fan arrangement around the rest of the several-leveled amphitheater. For the summit, seats were assigned by seniority. You had to be a four-star or civilian equivalent to sit at the head table. The 100th person in seniority that day was the Director of Naval Intelligence, a two-star. It was the second-most senior gathering of US government officials I saw in my life, and I was its point of contact for this one.

As an aside, the most senior gathering was at the Naval War College in 1985 when the chairman of the Joint Chief of Staff called a meeting of all four-star flag officers in the US military. I had overseen setting up the communications for that meeting, and all had not gone well. The communications links I needed to support the proceedings were very highly classified, and the manager of them was a sergeant who would not tell me his service, or his clearance level even though we were talking on a secure, encrypted phone.

"My service is unimportant. I need justification why you want these links," he replied when I asked him for the links and then his service.

The fact was, the conference itself and who was attending was at an even higher classification than his communication links. So I could not tell him. Nor would he tell me who his boss was. If I had that name I could check the database to see if his boss had the needed level of clearance.

I still have dreams about the hassles that ensued. We talked several times, and he was not helpful at all. I finally resolved the issue by informing the sergeant of who was coming indirectly. The Chief of Naval Operations' office forwarded a straight Top Secret (no code words, no cover words) message to my office. It was from the Chairman of the Joint Chiefs of Staff and requested agenda topics for the meeting. It had only six addresses, the heads of every military branch in the US armed services—and the White House. I forwarded that message to the sergeant in charge of the communications links. I told him to look at the addressees and think about it for a minute or two. I said that if I did not get my links immediately, I would send a message to those six men, the heads of our armed services, and the President, outlining his non-responsiveness. And I would be sure to highlight his name and point out he would not even identify his service to me or who his superior was. I got access to those links the next day.

Then the links went down during the meeting. The Navy chief in charge of restoring them at the satellite downlink site went home when his shift ended without telling his relief of the critical need to get them repaired as soon as possible.

Ten years before, I was stationed in Japan with a captain who was now the two-star admiral in charge of naval communications. In Japan he had relieved me as the Episcopal lay person overseeing the administrative side of the church. So I knew him well enough to know his nickname and his home address. We exchanged Christmas cards. I made it very clear to the maintenance chief on the other end of the line that if I did not get that link back up within an hour, I would call the admiral at his home, using his nickname. At that point, it was just after 10 PM. I had my links back online well before 11. That was the only time that I can recall ever being a complete hardass in my career. But I digress. Back to the planning for the MDA Summit.

The lab had its trained rapporteurs. Indeed, I had trained to be one when I first got to APL nearly 10 years earlier. This time I assumed the role of their coordinator, just as I had for the wargames at the Naval War College following 9/11. The upshot of the high level meeting was the decision to establish a very senior steering committee to oversee the writing of the National Strategy for Maritime Security. There were to be eight supporting plans as annexes, with the understanding they would be parsed out to the appropriate department and agencies by the White House. The resulting documents would define all members of the US government's roles in reference to its maritime security and maritime awareness.

Our office oversaw overall coordination for the plan, but we had a lot of "help," a good bit of it counterproductive. Everyone in the US government seemed to think they were in charge, or at least had the most important part of effort. It made for the busiest 15 months of my life. I worked over 12 hours every day for much of that time. Longer, if you count the time I spent on my smartphone on the train between DC and Baltimore five days a week.

I was assigned to write the first drafts of the National Strategy for Maritime Security and the National Plan to Achieve Maritime Domain Awareness. It was really just a slight reformatting of the paper I had been working on since late 2001. I was also asked to help write several others, including a smaller role in the Global Maritime Intelligence Integration Plan, the Interim Maritime Operational Threat Response Plan, and a more substantial role in the National Concept of Operations for Maritime Domain Awareness and the International Outreach and Coordination Strategy. Our office coordinated the many folks working on these plans all over the government. That made for a very hectic year for many of us, but we got it done. It was a genuine team effort, and I was proud to be part of it.

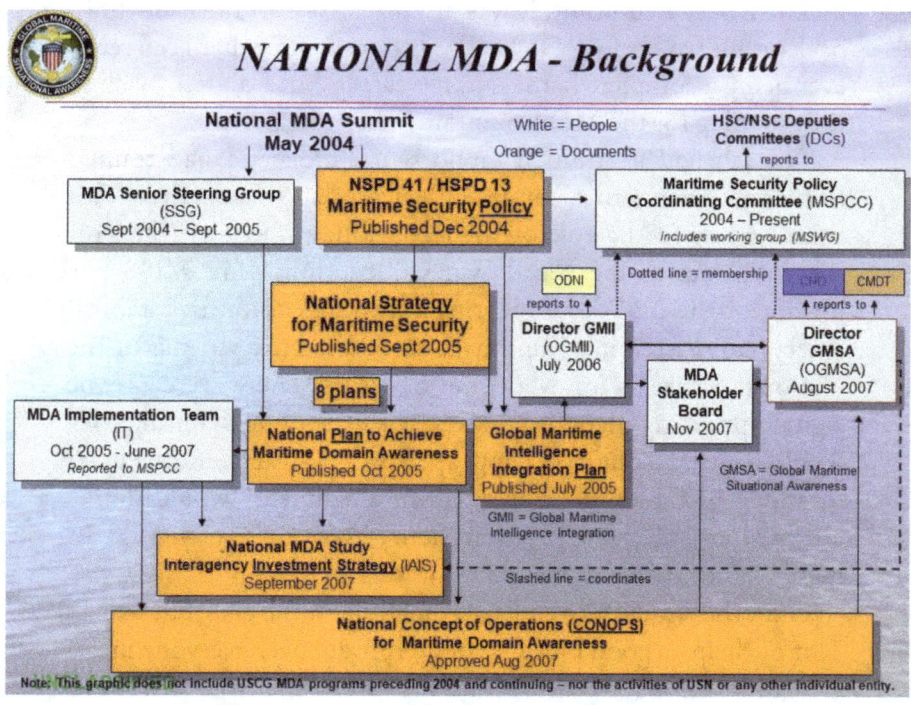

Schematic from National Office of Global Maritime
Situational Awareness Overview. May 2008. Courtesy of the author

A fair bit of what I initially wrote ended up on the cutting room floor, as they say in the movie business. Those first drafts had been meant as thought starters rather than finished products, and enough remained that I was very satisfied.

The statement issued on 1 September 2005 as the introduction to the National Strategy for Maritime Security says it all:

> *The safety and economic security of the United States depend upon the secure use of the world's oceans. Since the attacks of 11 September 2001, the Federal government has reviewed and strengthened all its strategies to combat the evolving threat in the War on Terrorism. Various departments have each carried out maritime security strategies which have provided an effective layer of security since 2001. In December 2004, the President directed the Secretaries of the Department of Defense and Homeland Security to lead the Federal effort to develop a comprehensive National Strategy for Maritime Security, to better integrate and synchronize the existing Department-level strategies and ensure their effective and efficient implementation.*

Maritime security is best achieved by blending public and private maritime security activities on a global scale into an integrated effort that addresses all maritime threats. The new National Strategy for Maritime Security aligns all Federal government maritime security programs and initiatives into a comprehensive and cohesive national effort involving appropriate Federal, State, local, and private sector entities.

In addition to the overall strategy, at one point there were eight supporting plans to address the specific threats and challenges of the maritime environment. While the plans address different aspects of maritime security, they are mutually linked and reinforce each other.

The supporting plans include:
- —National Plan to Achieve Domain Awareness
- —Global Maritime Intelligence Integration Plan
- —Interim Maritime Operational Threat Response Plan
- —International Outreach and Coordination Strategy
- —Maritime Infrastructure Recovery Plan
- —Maritime Transportation System Security Plan
- —Maritime Commerce Security Plan
- —Domestic Outreach Plan

Development of these plans was guided by the security principles outlined in this National Strategy for Maritime Security. These plans were to be updated on a periodic basis in response to changes in the maritime threat, the world environment, and national security policies.

Together, the National Strategy for Maritime Security and its eight supporting plans presented the first comprehensive national effort to promote global economic stability and protect legitimate activities while preventing hostile or illegal acts within the maritime domain.

Chapter Five

In Secret No More

Over the next several years, I worked on many tasks and participated in numerous studies, including co-chairing the development of the National Maritime Domain Awareness Technology Roadmap, which reported out in April 2005, and its partner, the Data Fusion Study, which we finished about six months later. I also worked on the Department of Homeland Security's "Wide Area Surveillance Study" of 2004, which recommended looking at aerostats and Predator-class drones as the best way to conduct active patrols out to 110 miles offshore.

All three were unique pieces of work that I referred to for the rest of my time in the government. All spawned follow-on studies that I believe were just baby steps after the original three documents I co-authored had catapulted knowledge in this arena forward by miles. But I may be a bit parochial here. I use the graphics we developed for the technology roadmap even today. I used my systems engineering textbook from my graduate studies at Johns Hopkins University as the blueprint for organizing that study, and I reached all the way back to my time as a research assistant at the Naval War College's Center for Advanced Research, where I had created the Warfare in the Fourth Dimension concept.

In all three marine domain awareness studies, we harked back to the MDA/PIO War Room and built a series of matrices listing every system we knew in each area: Platforms, Sensors, Processing/Fusion, Display/Decision Aids and Dissemination/Communications. We tried to keep it unclassified, but some team members, which had nearly 100 people, were from the intelligence community and wanted it classified due to the aggregation of data and knowledge. I had no argument to counter that, and so the original "Technology Roadmap" was marked *Secret* with an annex that dealt with national space systems classified significantly higher.

By five months after the release of the Technology Roadmap I had learned to live with a document that we needed to share with many organizations that did not have any clearances. We had excerpted parts to make our point to the various uncleared stakeholders when, in September of 2005, the National Security Agency forwarded an unclassified version of our report, which their security censors had blessed. It cut our original 150-page report about in half. But it was now something we could use to work with the many uncleared clients with whom we were dealing. Well done, NSA! We immediately sent it to many people.

About two weeks or so after the NSA gave us its supposedly unclassified report, there was a humorous event. One of the middle pages of the roadmap bore a drawing with high classification markings. The picture was significantly over-classified. Indeed, it was unclassified unless you knew the source, but the markings themselves were classified, and thus they gave a clear hint to the source. (Highly classified space systems.)

All of us had missed the classified markings, and now there were copies in many unclassified places. The Special Security Officers went berserk. This was about the same time some of the same folks were trying to retro-classify satellite AIS. I just sat in a corner and kept my mouth shut, but I was laughing inside, and I was extremely glad I was not the one that had initially cleared it for distribution.

The Interagency Investment Strategy was a different case. The man from our office who was in charge of organizing that effort did not seem to have a clue as to what he was doing. The people he was allegedly leading were very unimpressed, and I ended up ghost rewriting much of that document, too. I must have done an excellent job because our mutual bosses (there were two) liked the finished product so much they gave him a promotion! At the same time I was informed I was going to be "promoted" to the office of the vice commandant of the US Coast Guard to be his science advisor.

As is easy to imagine, I was incensed, and I very nearly resigned at that point. I went to see our boss and asked why I had not been given the job the non-performer had gotten. They pointed out that the Office of Vice-Commandant was a significant step up (three stars versus one in the rank structure). Besides, the Office Global Maritime Situational Awareness would be an office focused on policy, not an organization dealing with technology.

I countered by asking him how many times over the past years that we had been working together had I been invited into a meeting allegedly focused on policy after it had started. My unplanned inclusion was because the discussion revealed the policy in question hinged on what was possible technically. We both knew the answer was many times, certainly more than once a week if you count the follow-on meetings, which often did focus on technology.

I was somewhat mollified, but I noted that I had spent the last six years working on the Maritime Domain Awareness problem and that I was wedded to it in many ways. I had even resigned from an excellent job at Johns Hopkins to work on it. We reached a compromise: I was assigned to be the science and technology advisor for the new office, reporting directly to the new director, Navy Rear Admiral (lower half) Lee Metcalf. It was probably a professional career mistake. I should have gone to work for the vice admiral. At a minimum, I should have asked for the title of chief technical officer, as advisors are much easier to ignore.

On another unforgettable and uncomfortable note, not long after the National Strategy for Maritime Security was published, I happened to meet a retired Coast Guard officer who I had heard was one of the brightest minds in the Coast Guard. I had looked forward to meeting him.

Inexplicably, he was overtly hostile, even rude. I was shocked but didn't see where I could do anything to alleviate his belligerence, as I had no idea why he obviously felt so hostile. So I just went on my way.

However, a couple of years later, when the Secretary of the Navy created his Maritime Domain Awareness office, this man became one of its top leaders, and I had to deal with him. But his inappropriate actions toward me intensified rather than diminished, and I still had no idea why. It was as if I had personally insulted him somehow. About four years later, I discovered why when he was speaking at a national conference in San Diego on maritime security matters. I was in the audience and commented to the man next to me that it was undoubtedly an excellent speech. My neighbor, a well-known author on maritime and national security issues, replied, "Yes, indeed! He wrote the National Strategy for Maritime Security!"

Say what? I thought I had! At least the first complete draft. Jeff High, my boss at the time our team wrote the strategy, was also in the audience, and I sought him out immediately after the session ended.

"Jeff, Did my antagonist have anything to do with the writing of the NSMS?"

"Guy, he sent in an unsolicited draft of his ideas, but your draft had already covered every one of his points and was a better paper, so I just stuck it in my drawer and later pitched it."

My draft was the result and distillation of four years of effort, with input from literally hundreds of people and many, many revisions, so I was not surprised to learn mine was the better paper.

But now I know why my mistaken colleague had been so hostile. He was either jealous and/or believed I stole his paper and rewrote his ideas. Not true. I never knew his document existed until I spoke to Jeff. But he told others he had written the strategy, which was plain untrue. He was probably just embarrassed. That antagonism towards me later had severe national and international consequences with regard to Implementation Task #1 of the National Space Policy, released in June 2010—which I also wrote. It is easy to see the results of his revenge even today because that task was never implemented even though there is an obvious significant need for it.

He was the chairman of the national committee tasked to implement key elements of the National Space Policy on June 28, 2010. From September 2009 to June 2010, I had been an active member of the interagency team assembled to write a new National Space Policy at the direction of President Barack Obama. I was selected for that team by the Department of State lead of that team because he recognized the first page of White House tasking letter to the cabinet members was strikingly similar the first page of a "thought-starter" paper I had written several years before describing why we needed C-SIGMA and how to implement it. Task #1, which basically said "implement C-SIGMA," was the upshot of my very successful efforts to sell C-SIGMA to the White House committee, and at the direction of that committee, I had been laying the groundwork for its implementation starting in late October 2009.

In late June 2010 the maritime coordinating committee was tasked to write the actual implementation plan for that specific task, with the understanding at the White House that I would lead that effort. Shortly thereafter my mistaken gentleman, who happened to be the chairman of that committee, fired me in an insulting and abrupt way, during a meeting of the committee as the person who was going to lead the implementation effort. Indeed, that task and thus C-SIGMA was, much to my regret, never implemented. The mistaken gentleman had his revenge for the supposed slight, but the United States lost a chance to be the global leader in aligning the space age with the maritime world to the betterment of all on Earth.

In mid-2022 the QUAD, a consortium of US, Japan, India and Australia, all countries I had visited and spoken on C-SIGMA at dozens of conferences there, as well as over most of the world, in the past dozen years, enacted an agreement that they would band together to conduct surveillance of the Chinese fishing fleet primarily using unclassified space systems. It is a step in the right direction, but covers only one of the many problems we face at sea that could be alleviated by implementing C-SIGMA. The QUAD missed a golden opportunity.

I was going to quote National Space Policy Task #1 verbatim in this book as it is marked with a *U* in the original implementation directive, which means it is unclassified and thus can be quoted. However, of the 305 pages I submitted for security review (which took 19 months), the only paragraph that was ordered deleted in toto was the quote of that unclassified paragraph. Very, very strange! Why is the wording of this now superseded, unclassified task being hidden from the public? I will discuss this further in Chapter Ten on C-SIGMA.

Moving on

Jeff High retired in May 2005, and Coast Guard Rear Admiral (lower half) Joseph Nimmich was named the head of the Maritime Domain Awareness Program Integration Office. One of the first things he did was to have everyone write a one-page paper on what they were doing for the office. After he had read the papers, he had each one of us in to discuss their work with him, one on one.

I gave a short description of my responsibilities and then a list of 18 different projects, in priority order, on which I was working. About in the middle of my list was ACTDs/JCTDs and he asked what that meant. I described the Department of Defense Advanced Concept Technology Demonstration program. I explained this was the process by which the DOD examines, field tests, and advances the state of the art of literally all technologies in which they were interested.

I also mentioned that it had recently changed its name to Joint Capabilities Technology Demonstration to reiterate that it was very interested in "jointness." Because of this new focus on jointness, the Coast Guard and, indeed, all tactical units with the Department of Homeland Security, were being warmly welcomed.

I had been involved with ACTDs at Johns Hopkins before I went to the Navy Warfare Development Command in early 2000. This program was also a significant input to the Fleet Battle Experiments (FBEs) I had participated in at Newport from 1998 to 2003. I also explained how I had gotten the Coast Guard and Homeland Security deeply involved as partners and how important I believed it was. He agreed and told me to make it my number one priority, which I very willingly did.

I retired seven years later, and DOD gave me a beautiful plaque citing me for saving the Coast Guard "many millions of dollars in research and development costs by leveraging the Joint Capabilities Technology Demonstration process." I proudly placed the memento on my living room wall.

The plaque's inscription is an accurate description of what transpired over those years. Joe Nimmich saw the utility of the

Coast Guard getting more deeply involved in JCTDs. He handed the ball to me, and I ran with it.

Other than satellite AIS, I believe getting the Coast Guard and the Department of Homeland Security involved in the process and forcefully presenting both organizations' needs to the DOD test and acquisition communities was the most significant thing I did during my nine years with the Coast Guard.

Unfortunately, when I retired, the deputy chief of Coast Guard research and development killed his service's participation in the program. I was shocked to learn that, but I always knew he was basically an accountant and had no idea how R&D was done.

For instance, unless you genuinely have no idea of where you need to go, the situation we were in when we started the research on how to counter maritime terrorism, a requirements-to-capabilities roadmap, or blueprint, is a must. If it is the former case, you realize you do not know what you do not know. A wargame is an excellent place to start, to define your requirements and get some idea of what you do not know. I had gotten the USCG involved in some very pertinent wargames which were telling us a great deal where we should spend our R&D budget, but he killed our participation is this series of very pertinent wargames as well. As I said above, he had no concept how R&D was done. Test and evaluation, yes, R&D, no. I will describe why this is pertinent in a few pages, but let's return to the start of Admiral Nimmich's term before I return to the importance of wargames.

In addition to asking us to itemize the projects we were working on, Admiral Nimmich asked us to list who we were working with and indicate our main support person. Then he published it. It turned out the office was working on about 25 projects besides my 18, which were all external. And I was listed as the main person in support on 23 of the 25. That meant I was working on 41 tasks. No wonder I was so busy.

As I mentioned above, the Interagency Investment Strategy was a different case. The man we had leading that effort was the one truly weak sister we had in the office. All talk and show, but no leadership skills as near as most of us could see, and he made a habit of claiming other people's work for his own. The team members were from many different organizations, and I knew many of them from my earlier experiences. Word soon came back that the team was very unimpressed. Sotto voce, I ended up ghost rewriting much of that document, too. In retrospect, I should have had a private conference with Admiral Nimmich and discussed the situation, but I did not. My error. I just did not feel comfortable walking into the office of anyone and running down a fellow worker. I believed that it was evident that it would be detected soon enough. I was wrong there, too.

Admiral Nimmich moved on as the organization morphed into the Global Maritime Situational Awareness Office. Navy Rear Admiral Lee Metcalf, an activated reservist one-star admiral, was assigned to lead us. He had a very tough job ahead of him as the Secretary of the Navy had just set up his own organization with precisely the same task.

The Navy was creating several Senior Executive Service (SES) billets, billets we thought were going to come to us (with one of them, for me) on Metcalf's staff to do exactly the job we had been doing for four years in DC and two previous years in Newport. The Navy refused to fund our office. Then the leader of that office informed Admiral Metcalf that our office reported to him. Say what? We were, after over five years of discussions and planning, a White House-established office. The SecNav office leader was a political appointee in an office set up by the stroke of a pen by another political appointee. Neither had any experience in maritime security. National politics; yes, maritime awareness? None! Our boss, Lee Metcalf, was trapped. All admirals report to the Secretary of the Navy. His word is binding.

To say I was disappointed is an understatement. The Secretary of the Navy brought in a group of competent folks, but none had any experience with multinational marine domain awareness, the task assigned to Metcalf, or research and development, the task assigned to me. To justify their existence, the new people immediately began belittling our work, even though they obviously knew extraordinarily little about it.

In that these were the billets and money we thought we would get to implement the Maritime Domain Awareness Technology

Roadmap, published two years before—after four years of research—we became not just disappointed, but angry. Even angrier when the word started drifting back from several sources that the Secretary of the Navy folks were highly critical of our work and very vocal about it. It was evident to the folks reporting back to us that the new guys had no experience in this area.

Indeed, one of the Senior Executive Service members in the Secretary of the Navy's office went so far as to send an email to the Dean of the Naval Postgraduate School. It stated that the school could expect to not get another dollar from the Pentagon for Maritime Domain Awareness research if he did not stop working with our office and, specifically, me. This senior executive worked for the same official who had been so hostile to me, so I have a good idea where that direction originated. I still have the email detailing this. The dean had been a professional colleague of mine for years, stretching back to before I went to Newport in 2000. He forwarded it to me with apologies, saying it was the strangest email he had ever received. He also stressed he wished to keep working with me but that we would have to be covert in our dealings from there on out—and so we were until we both retired at about the same time in 2012.

Still more bureaucratic battles

A year before the OGMSA office was created, Rear Admiral Rich Kelly, the commander of Joint Interagency Task Force, was transferred to Washington to lead the Office of Global Maritime Intelligence Integration.

He was first in his class at the Coast Guard Academy, a cutterman (Coast Guard for a surface officer as opposed to an aviator), with a good bit of time at sea in both Coast Guard and Navy warships on smuggling interdiction patrols. And he was a Harvard-educated lawyer. The primary mission of the task force was drug interdiction, and Kelly had worked closely with all elements of the intelligence community, including Customs and the Drug Enforcement Agency. So he was the perfect person for this assignment. I immediately started assisting him and his new staff.

Thus, I had a good relationship with that office when Admiral Metcalf came onboard as director. He and Kelly, with members of both staffs, had the first of what became several coordination meetings. The first meeting, held in the Office of Global Maritime Situational Awareness, did not go well at all. Both admirals saw their jobs as entailing responsibilities which the other saw as his. It was all a bit sticky, and Kelly abruptly left with no resolution to any issues.

Several more meetings followed and eventually, agreements for mutual support were reached. I had a small role as a peacemaker as I had established cordial relationships with Kelly and his staff during the period before Metcalf arrived.

The two gentlemen, both professionals, eventually established a working relationship, and I was officially assigned to both organizations with the title United States Science and Technology Advisor for Maritime Domain Awareness. It was the highpoint of my professional life as far as titles go, but it did not last long. Both men were fired within 24 months (both for doing the job given them by the President via National Security Council).

The Navy secretary's MDA office took exception to my title as they saw themselves as having that responsibility, the signature of the president notwithstanding. Not to put too fine a point on it, but to my understanding, the president of the United States outranks an assistant deputy secretary of the Navy, and no one in the office had anything approaching my level of experience in the R&D world. So I just ignored the snide remarks being reported back to me. These were primarily fine naval officers with significant line experience with two retired Coast Guard officers, one a captain and the other a warrant. As near as I could tell, none of them, Navy or Coast Guard, had any R&D experience. Not any at all. They had never even heard of JCTDs or participated in the organization, execution, or analysis of a Fleet Battle Experiment. No one in the Secretary of the Navy's office initially knew the Joint Capability Technology Demonstration process even existed, much less its usefulness for getting research and development funds to examine operational problems—and its usefulness in solving many of those issues.

After eight months or so, they discovered it somehow and approached the JCTD office, asking how they could get involved. I was good friends with everyone in that office, having been working with it for about 10 years. My friends there told the Navy secretary's office that they should come to see me for assistance in this area as I had been working with them for many years in precisely the area in which they were interested. The office handled Army, Navy, Marine, and Air Force requirements, and I had been beating the Navy, and Coast Guard drum for many years at that point.

Both Navy personnel and people from the Secretary of Defense office later told me much the same story from two vastly different viewpoints. I swallowed my anger, allowing my sense of professionalism to rise above it and helped them understand the process and how to use it. Then I watched, I must admit, with great glee, as they chose their first three projects to fund at the end of their first 18 months of existence. All three were projects I had conceived and started or significantly helped to do so. In two cases, they changed the name, but for Regional Maritime Awareness Capability, they even kept the name. It was just too funny that they had belittled our work for 18 months, and then the first three projects they funded were ones we had created. I still chuckle when I think of it.

I will compete with anyone "nose up," as we used to say in drag racing, which I did my fair share of as a young man, but the Navy Secretary's office was not playing fair at all. They had us at a significant disadvantage. They were insulting our efforts of the past several years, but we could not insult theirs because they had not done anything yet except go to meetings and run their mouths.

I believe our national security still today has suffered from their actions. Our ports and coasts are still woefully unprotected. If the White House had implemented the Maritime Domain Awareness Technology Roadmap of 2005, the ports and much of our coasts would be much more secure today. I am sure the Secretary of the Navy thought he was protecting the interests of the Navy when he set up his shop, but I am equally sure his actions did not advance the security of the United States. To give the devil his due, I am also sure he did not mean to do that; he just had not thought the problem through to its most probable conclusion. There should never have been two Maritime Domain Awareness organizations. The Navy office was, in my opinion, a discredit to the Navy and the United States. It was the equivalent of assigning a bunch of fighter pilots to run the surgery department of a significant, very busy hospital. In that situation, the fighter pilots would probably run around telling everyone that they were the best in the world, but nothing would get done. So it was here, except the fighter pilots would be intelligent and self-aware enough to ask for assistance.

The SecNav MDA office sniping took its toll. As I watched the harassment Lee Metcalf was getting from the Navy secretary's office, I felt deeply sorry for him. He was new in this game while I had been in it basically since 1968, and very specifically since the late 1990s while working at the best naval research organization in the US, if not the world, Johns Hopkins University's Applied Physics Laboratory. I could take it as I recognized what neophytes, what babes in the woods, the Navy Secretary's Maritime Domain Awareness office were.

To be entirely fair, that office had some good folks, including all the Navy commanders, who were very professional, and really stood out from those who were not. Still, no one in that office had anything like my depth and breadth of experience, so I just shrugged off their baloney. On the other hand, Metcalf hoped to make another star, so he was sweating all the insults and slights.

The Deputy Under Secretary, a political appointee, had taken it upon himself to significantly modify the White House guidance given to our office. He ordered Metcalf not to talk to any area commanders (Pacific Command, Atlantic Command, European Command, etc.) or their staff.

Admiral Metcalf had tried to obey, but he could not do the job the White House had given him without talking to someone at these organizations. To do this task, he started dealing with the civilian Department of State advisors at each command. He believed the verbal guidance he had received dealt with just the military. The Deputy Under Secretary discovered this and summarily fired him, in the Secretary of the Navy's name.

The Chief of Naval Operations got involved and moved Metcalf to another billet, one not under the immediate jurisdiction of the political appointee. The Office of Global Maritime Situational Awareness was now leaderless. During his 14-month tenure, RMDL Metcalf repeatedly asked the Coast Guard to take back the person they had assigned as his deputy, a smooth-talking lightweight.

When this lackluster person did assume the acting leadership role, the organization was doomed. I never took the deputy role, and the Office of Global Maritime Situational Awareness drifted into irrelevance except in two areas—technology and outreach. We had some recalled reservists who had significant merchant marine backgrounds, and they were our interface to the civilian maritime world. Unfortunately, about 18 months after Metcalf was fired, the Navy stopped funding the activated reservists and sent them all home, at least in large part because our non-leadership had done almost nothing to defend them. I went all the way to the vice admiral level in both the Coast Guard and the Navy trying to save them. But it was much too late.

It was even worse, I gather, at the classified level. Admiral Kelly had held the job of directing the implementation of the Global Maritime Intelligence Integration Plan for just over two years when he became so frustrated late on a Friday afternoon that he said something very rude to his boss on the IC staff. He walked out the door, intending, he told me, to resign on Monday morning.

He did not get the chance. When he arrived at his office in a classified building on Monday, the guards informed him his clearance had been revoked. He was escorted to his office to pick up his personal belongings. The Coast Guard, his parent organization, gave him the office next to mine. He was there just long enough to process his retirement papers.

I greeted him the day he arrived, and he told me the story above, but I did not realize how short that time was going to be. I recall he was given about two months, maybe even less, to get his affairs in order and retire.

A very few weeks later, late one Friday afternoon, I walked into his office to ask him the date and place of his retirement ceremony.

He told me that there was not going to be any ceremony. He was still angry at the way he had been treated. At the end of that day, in just a few minutes, he would be walking out the door and might never wear his uniform ever again, nor set foot in a government facility by choice. In that he and I were the last two in the office that day, I believe I was the last person with whom he had a conversation while in uniform.

Shortly after that the United States' Maritime Domain Awareness function moved from the Coast Guard headquarters, which was open to anyone with an escort, to the Office of Naval Intelligence complex at Suitland, Maryland, which required an SCI (Sensitive Compartmented Information) clearance just to pass the lobby. With the placement of MDA behind the "Green Door" and into the SCI world, MDA, as defined by both the concept paper of the 1990s, and with our efforts starting in 2001, died.

Outreach to the civilian maritime world and most foreign nations became just too hard, and I transferred to the full-time position as the Science & Technology Advisor for the Coast Guard. I considered it a demotion, but it was the only path forward I could see. Besides, I was still an advisor to the Maritime Domain Awareness office in at ONI Headquarters in Suitland on a case-by-case basis. But they were now almost entirely focused on policy, not technology.

I did spend a good bit of time there. I made a point of telling anyone who would listen—including all three of the admirals that led the new organization in rapid succession—that the MDA shop needed its own science and technology advisor. I did not want the job as I was now focused on unclassified space systems.

I understand that as of mid-2019, the MDA office has restarted its outreach program and is working diligently to interface with the commercial world. Good for them! About time.

Chapter Six

In the Realm of Space, a Nod From Rahm

In early 2006, I wrote a paper called "International Collaboration Is THE Silver Bullet" about the future utility of the new space systems scheduled to come online in the next few years. In it, I stressed that collaboration among all the world's nations to build a space-based maritime awareness system would build mutual trust and friendship, dramatically improve maritime security, safety, and protection of the maritime environment and its resources.

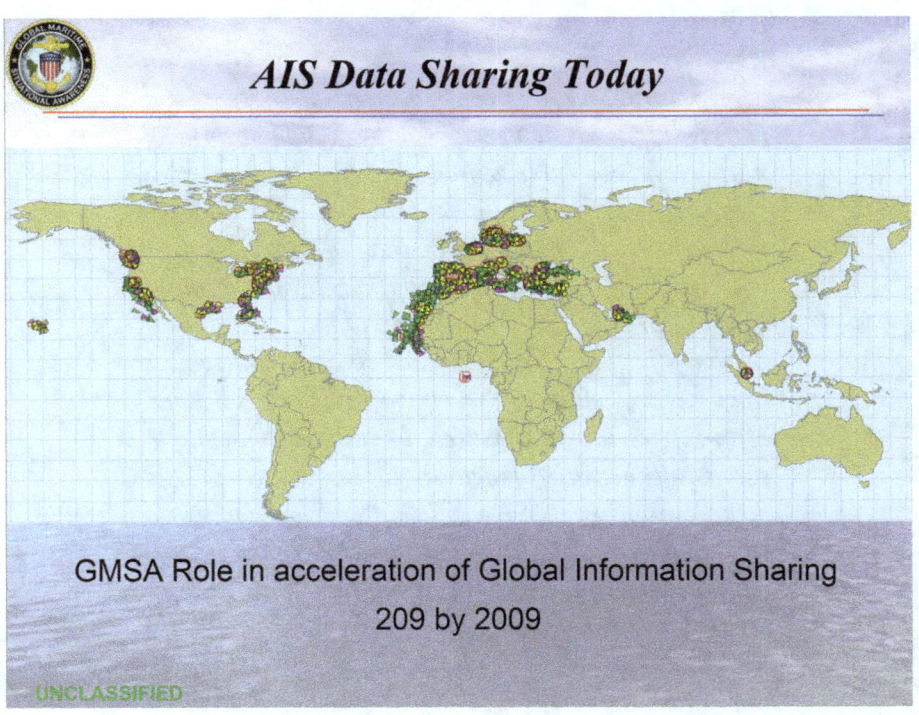

Slide graphic from National Office of Global Maritime Situational Awareness Overview. May 2008. Courtesy of the author

I had sent it to many people, and by 2009 bits and pieces had been published here and there. One of the people I had sent it to was a high-ranking State Department official. In late July/early August 2009, he gave me a call.

"Guy, are you aware you have been plagiarized by the Chief of Staff of the president of the United States?"

"No, but I am flattered, not offended," I replied. "Tell me more!"

It turned out that President Barack Obama, seven months into the office, had waxed eloquent at a recent cabinet meeting on just the same thoughts regarding using Space to build friendship and understanding globally as were expressed in my three-year-old paper. He had ordered a rewrite of our national space policy, and Rahm Emanuel, his Chief of Staff (CoS), had drafted a tasking letter for all departments and agencies. His opening paragraph sure looked a lot like the opening paragraph of my original. Coincidence? Maybe, maybe not. I never found out. This is what I had written:

> *The potential unique contributions of current and planned space systems, owned by a wide range of nations and available to many others, to international global maritime awareness is a subject of growing interest to many. However, in order to understand the true uniqueness of those contributions the background needs to be set first. Many individuals and organizations that have closely studied the problem realize that no one country or even any existing collection of countries has the stature, breadth and depth to successfully organize a meaningful coalition to protect oceanic commerce, the maritime environment, and the broad range of individuals that use the maritime domain for a multitude of endeavors including profit, conveyance, and recreation. They realize it will take international collaboration and cooperation on an unparalleled scale to provide this protection and assure the safe and secure use of the world's oceans. The only organization that has addressed a task similar in scope*

is the International Civil Aviation Organization (ICAO) and that effort took almost 40 years to reach full functionality after the need was first articulated. Many believe that because the maritime domain has been an integral part of the world's commerce and conveyance systems for thousands of years it will be much harder to create the needed organization to regulate it.

The State Department was leading the writing of the outreach portion of our National Space Policy. In that I seemed to be right on the same wavelength as the president, and had obviously done a good bit of research in this area, they asked if I was available for a day a week for the next six to nine months to work on the drafting of a new national space policy. I told him I was very interested in joining their effort, but I had better check with my chain of command and would get right back with him. In that I had participated in drafting both our National Strategy for Maritime Security and several other related documents, I did know the drill. (It is a lot like watching sausage being made.)

I immediately asked our lead in the drifting Office of Global Maritime Situational Awareness, stressing that I would be out of the office one day a week on White House business. He replied: "Whatever. Do it if you want to," in his classic non-leadership style. I am certain he was very displeased that I had been asked, by name, to do this job, as he went out of his way to suck up to influential people or those he thought were important. The fact that I, not him, was being asked to participate in this White House working group wholly frosted him, but he could not see how he could turn the request down without causing a stir.

Between September 2009 and 21 June 2010, I spent about a day a week attending meetings at many different departments, including State, Commerce, Defense, Transportation and Homeland Security. Several intelligence community elements and the Pentagon also hosted us and we met several times at the beautiful Old Executive Office Building adjacent to the White House.

In late October, the team members were all asked to write a paragraph or more on what each of us, individually, believed should be included in the new National Space Policy.

"Write the words as they should appear in the NSP, and then a page or less on the pros and cons," were our instructions.

Fifty-one people, including myself, did as requested and submitted proposed wording for the new national policy.

White House staff discussed our work and ranked them from 1 to 51. Rahm Emanuel, as the drafting effort leader, was directly involved in the selection and ranking. The diminutive, excitable Emanuel was known both for his political and policy skills, coming from Congress to join the administration of fellow Chicagoan Obama. He is a real player, elected later as mayor of Chicago and even becoming ambassador to Japan. The following week I learned that my input had been ranked Number One for innovation and utility, and that Emanuel liked it very much. I did a little jig!

Over the next seven-plus months, I also spent significant time at meetings in and with the State Department. They even had me brief the C-SIGMA concept at several conferences, including one with the European Space Agency, where C-SIGMA received a warm reception. The European Space Agency invited me to Frascati, Italy, the home of the agency's Earth Observation Research Center, (Beautiful place! Wonderful wines!) where we subsequently held our first C-SIGMA conference.

Over this entire time I was working on the draft National Space Policy. I included my boss on all significant emails, had shown him my drafts, and asked for his comments or input, none of which was forthcoming. No surprise. I also briefed the progress we were making on this project at the weekly staff meetings, which saved my bacon and cooked his.

The White House forwarded the draft National Space Policy to cabinet members and departmental leaders for comment. It was precisely 300 lines long. Forty-five of those lines, by count, were mine. Those 45 lines dealt with both the need and the processes required to align our space policy with our maritime strategy for the good of the United States and all people's benefit on Earth.

When Homeland Security Secretary Janet Napolitano received her copy, she immediately recognized that a lot of what I had said would impact the US Coast Guard, which was in her department. Indeed, that was the main reason I had kept my chain of command at Coast Guard Headquarters informed.

I gather she picked up the phone and asked the commandant, Admiral Thad Allen, if he had input to the new policy for space and did he know who had written these words. I understand that Allen gave the standard junior officer to the senior officer reply. "I do not know, but I will find out." He tasked his head of policy to find out who had written these words.

The commandant has a meeting for all department heads every Tuesday that my boss was required to attend and to update the senior staff. I had assumed my boss was passing on my weekly updates in those staff meetings, as it involved the Coast Guard. I was telling him very clearly that I was helping to write a national policy that would affect the Coast Guard. It never dawned on me that he would not keep leadership informed. The only possible reason that I can see is that he was so very jealous of me that he did not want the senior leadership to know what I was doing, especially that I was working very closely with the White House staff. In hindsight, I should have asked him if he was getting any feedback, but I had assumed he would tell me if he had.

Because my boss had not told anyone about my involvement with the White House, the Coast Guard Director of Policy could not find anyone in headquarters who knew anything about the draft NSP. The gentleman had his deputy, a captain, phone the White House staff, who gave them my name and told them I had been a part of the team and a visitor at the White House off and on for the past seven-plus months.

The first I knew about all this was when the head of policy, a very senior civilian, and his captain deputy were in front of my desk doing a war dance, threatening to string me up from the yardarm of the flagpole or at least get me fired. THAT DAY.

I was shocked that they did not know about it. But then I considered the lack of character of my boss and guessed what had happened. However, I was not too worried. I knew I had kept my boss informed, and I had a long string of emails detailing what I had told him and when, to prove it. I told the two men that. "Where is his office?" was their next question. I told them it was two doors down to the left.

BANG! The two gentlemen flew out of my office, performed a full speed, hard-left turn, and stormed down the hall into the boss's office. He was meeting with two of our captains, one from the Coast Guard and one from the Navy, both friends and colleagues of mine at that time. The two gentlemen from policy did not even wait until the room cleared before asking him to explain himself for not passing on information about my national-level task.

My boss denied all knowledge of what I had been doing. This lie was in front of my two colleagues, the two captains from our office. Both came to tell me our boss had just lied to the two policy gentlemen, as they had been present numerous times when I had briefed him at staff meetings on what I was doing on the space policy project.

Shortly after, our so-called boss disappeared to work as a shift worker at Homeland Security headquarters, a huge demotion. I understand he was subsequently ordered to Djibouti as a watch-stander there. A short time later, the White House disbanded our office. Parts of it went to work at the new MDA office being set up at the Office of Naval Intelligence. The rest of us found other jobs.

Our outreach mission, which was probably both the most unique and the most vital, and influential part of our organization, when measured for bang for the buck and overall effect, was disbanded, much to my dismay. I think the whole office felt that way. They were the only ones, besides me, who were clearly pulling their share.

Indeed, they were punching well above their weight, and I was proud to be a small part of their effort, assisting them where I could. I was, obviously, proud of my work, too, but theirs was unique and extremely valuable in the big scheme of national and international maritime security.

Responsibility for the furtherance of MDA moved over to Naval Intelligence, a great organization with many fine people. However, the outreach to the merchant and civil maritime world is a very tough row to hoe for an intelligence organization. Classification problems and distrust on both sides just keep slowing down progress in the maritime domain, if not canceling it outright.

I moved over to Coast Guard Research & Development, but it was not the same. They were a friendly group of folks in my new office, and the captain heading that office was a peach. But they were working at a level well below what I was used to working. I had been working with the leadership of the Office of Naval Research. They worked with mid-grade officers at best.

I am sure I brought in more funds for technology advancement, especially in the C4ISR (Command, Control, Communications, Computers (C4) Intelligence, Surveillance and Reconnaissance) realm via the Joint Capability Technology Demonstration program, than did the entire rest of the 20+ man R&D organization. They were just not engaged in true R&D, but rather in test and evaluation (T&E) of systems on which others had done the research and development.

Most of the Navy and Defense Department money I assisted in getting spent on Coast Guard and Homeland Security needs went directly to the operating forces, so the headquarters never really understood what was happening, except for a few folks. The fact that the headquarters did not know or understand what I was doing was another problem from us not having a true leader from October 2009 to November 2011, when the White House disbanded the office.

Part of it was all my fault, but I thought it would be unseemly for me to crow about my successes, which I thought were self-evident. I now realize the Coast Guard did not recognize my achievements because my boss had been hiding them due to jealousy. I should have been more attentive to my home front.

When R&D stands for retard and dispute

The R&D office clearly did not understand the entire process from its beginnings through test and evaluation of ready-to-be-fielded systems. They were involved in the T&E of finished products and never gave any serious thought about where all the items they were evaluating came from.

A classic example of what I am talking about happened about six months before I retired, when a company in northern Virginia, less than a 25-minute metro ride from our office, asked me to come to see a demo of a new signal processor. This new technology reportedly gave an inexpensive, moderate resolution analog radar the capabilities of an expensive high-resolution digital radar at a fraction of the cost.

I told the deputy of R&D that I was going, and he told me not to go, as it was not my job to explore new technology. Say what? That was exactly what my job was and had been for the eight years I had been there. Everyone else in the organization was doing T&E, but my job, by definition, was to do Research & Development (and Innovation). It seemed to be a foreign concept to him and nearly everyone else in the organization.

I was intimately familiar with the long tail of how a system comes into being because it was what I had been doing basically since I was hired at E-Systems in 1989, 22 years before. I tried numerous times to explain all this, but I might as well have been speaking Chinese.

Another example of complete non-understanding was when the Naval War College asked me to help lead a wargame at Newport on Arctic operations. As part of my past MDA duties, I had also been looking at the Arctic operations problem for some years and had even suggested to my ex-colleagues at the Naval War College that they stage a wargame to look at the problem; to find out what they did not know they did not know. The wargame was my concept, and I had helped set it up and generate a list of folks who should participate. I included myself as a subject matter expert on unclassified space systems. By then, I had been studying them for several years, and it was clear these systems were going to play a significant role in Arctic situational awareness. However, the office turned down the request from Newport.

"It is not your job to play in wargames," I was told.

Another example: At the European Command Operational Requirements/Joint Concept Technology Demonstration conference the previous year, I'd given a talk on how and why commercial space was going to be especially useful in Arctic operations. The European Command had liked my brief at several levels. They asked me to chair their annual Arctic Operations and Technology Day for both of the next two years.

The answer came back: "It is not your job to chair meetings on Arctic Operations." These were meetings about priorities in spending on R&D in preparation to operate in the Arctic, which would involve the Coast Guard. But that was utterly foreign to the folks in R&D.

I was also officially censured by the security office for saying that "commercial space systems could be a significant assistance to our NTM" (the unclassified way to refer to our systems used for national technical means of treaty verification.) The charge was complete and total nonsense. Even then, it was effortless finding official US government references to exactly this all over the web. One two-year-old unclassified congressional report directed the setting up of a panel to study precisely this issue. Attached was a 74-page report, the response to that order.

The report went into substantial detail about how this could be done, something I had not even addressed. Four or five retired three or four-star generals and admirals had signed the report. If my alleged violation of national security deserved a bad mark in my record, they needed to go to jail. However, the truth was, none of us violated anything. It was just ignorance on the part of Coast Guard security people.

There were many things wrong there, but I had neither the power nor the will at that point to tilt at those windmills. In the middle of all this, a Government Accounting Office report came out saying that the Coast Guard had 1,400 people in the DC area beyond their authorization. The GAO recommended cutting all 1,400 people. That is a huge number when you realize the total Coast Guard presence in DC was less than 4,500. It was a one third reduction in manpower. An edict went out that if you were not in an authorized billet that directly supported current operations or personnel management, you needed to find a new job.

That mandate was a double hit on me. My billet was one of those that had never been officially approved. And, being involved in R&D, I was most definitely not involved with current operations. I was offered a GS-15 job as head of approval authority for certifying authority to install and authority to operate electronic equipment. It was basically an accounting job and sounded dead boring, so I decided it was time to retire to concentrate on C-SIGMA, the child of S-AIS.

My retirement ceremony was very average until the very end when I was thanking the guests. I noted there were 18 foreign nations present. At that point, Commodore Ranjit Rai, retired from the Indian Navy, raised his hand from the audience and asked to be allowed to address the guests. I, of course, said, "Of course," and identified him as a retired naval flag officer from India with whom I had shared a podium on several occasions.

Rai noted that he'd been director of India's Naval Intelligence and later his service's head of operations. (His credentials didn't matter to some; the MDA office at the Office of Naval Intelligence, refused to admit him, saying it was inappropriate to let a retired foreign officer into its spaces without a formal government-to-government request.) Rai remarked that I was very well known in India, and that he knew I was also well known in Japan and Europe, too. He went on to say that the Indian Navy now had systems on all their ships that were the result of my writing and research.

It was the most remarkable thing I had ever seen, or even heard of, at a retirement ceremony. It also reminded the master of ceremonies that the United States had an award for me, too. He presented me with the Department of Homeland Security's Distinguished Career Service Award, the highest award given by DHS. I had never even heard of this award, and I am told that at the time, it was the first to be awarded in four years.

All's well that ends well.

Chapter Seven

When Wargames Expose the Vulnerable

*Something incredibly significant
to the Navy, the nation, and the world*

Many things changed after 9/11, and my life was one of them. The morning after that terrible day, 12 September 2001, I had arrived at my office at the Naval War College's Center for Naval Warfare Studies, wondering if my life and the lives of my many friends in or associated with the military were going to change much because of the horrific terrorist attack the day before. I suspected the change would be significant, but I had no idea how meaningful it would be.

Dr. Dan Smyth, the Center for Naval Analysis representative to the Naval War College and I, the Johns Hopkins Applied Physics Lab's representative, shared a converted classroom as an office at the Center for Warfare Studies with three other men who worked as senior analysts/subject matter experts in a range of fields: anti-submarine warfare; surface warfare; and anti-air warfare among them. A senior Naval Criminal Investigative Service agent was also there with us.

Dan was usually there at 0630, well before any of us. However, on the day after 9/11, he was not in the office when I got there, and I wondered why. He and I had left the building together the day before. We were all sent home early to look after our families on that dreadful day. The Naval War College leadership was concerned that terrorists might consider the college a high-value target, and none knew what would happen next that apprehensive afternoon.

As we walked out of the building and headed home, Dan and I had agreed that the United States' maritime assets would make high value, high payoff targets for terrorism. His first words when he finally walked into our office at about 0805 that next morning were: "Do you remember what we talked about as we left yesterday?"

Without waiting for a reply, he went on: "Well, the same thought occurred to the president, and yesterday afternoon he tasked the Chief of Naval Operations, Admiral Vern Clark, to conduct a study into the vulnerability of our maritime assets and develop a plan to protect them." He continued: "The CNO has tasked the president of the Naval War College, and late yesterday he called me at home to task me to lead the response to President Bush. I have just come from his office, and we are to immediately organize three wargames to define the problem and develop a strawman concept of operations."

He added: "Guy, with your wargaming and surveillance systems experience, I will need you to be my principal deputy." He paused to take a breath. He was excited, and upon hearing his words, so was I.

It took me a few minutes to get my head around the fact that Dan Smyth, my officemate, had just been officially tasked via the Navy's chain of command to respond to President Bush's order to the head of the Navy. The Navy had been ordered to identify the nation's vulnerabilities to maritime terrorists and to develop a plan to counter and mitigate them, and I had just been given a critical role in the response.

The Chief of Naval Operations decided that the best way to approach this problem would be via a series of wargames attended by the stakeholders of the maritime world. He recognized that to do a vulnerability analysis of the maritime assets of the United States, we were going to need input from everyone with a stake in it. This included governmental agencies from the federal down to the city level, and many private entities, such as shipyards, port and harbor managers, shipbuilders, brokers and suppliers, private and governmental, civilian, and military.

We were to work in close conjunction with the War Gaming Department of the Center for Naval Warfare Studies, but we were the lead. And I was lead on developing the required wargame program. Dan and I set to organizing the first game, which was 12 days away. I immediately headed over to the Wargaming Center across the street out our backdoor in McCarty Little Hall to start designing the game. Dan got on the phone to begin gathering names on who should attend and sending out invitations. He had the more challenging job. The list of stakeholders in the maritime world is exceptionally long, and he tried to invite at least senior representatives, if not the leader, from all of them.

The college's Wargaming Department is well versed in holding complicated wargames with a range of classification levels simultaneously, from entirely unclassified to highly compartmented intelligence. Indeed, while assigned there from to 1982-1986 I had created a pair of cells at different compartmented intelligence levels. One cell was tasked to insert an accurate portrayal of special operations forces' capabilities into wargames as required. The other cell was designed to add all aspects of satellite capabilities at the right level of technology for the game year.

By the time I left the Naval War College in April 1986 I had almost five years of experience with wargaming. By 2001 I had over eight additional years of experience at other venues, including two years at what is now the Joint Electromagnetic Warfare Center, reporting directly to the Joint Chiefs of Staff, and another year later at what is now the Naval Air Research Center at Naval Air Station Patuxent River in southern Maryland. Finally, and most importantly, I had five-plus years at the Warfare Analysis Lab at Johns Hopkins Applied Physics Lab, the world leader in multifaceted wargames, stretching from pure research through applied research to tactics, operations, strategy, and policy.

In the next several months, my substantial experience at the two leading institutions of wargaming in the US, the Warfare Analysis Lab at APL and at the Naval War College's War Gaming Center, would be expanded even further. The War Gaming Center was established in the 1880s and is renowned as a significant innovator. The wargames envisioned immediately after 9/11 were designed to bring together all maritime elements of the US and our main allies. They also included all stakeholders in both the public and private sectors, including all branches of the US, state and local governments, and all associated private parties such as ship, shipyards, port and harbor owners, builders, and operators together to highlight known vulnerabilities and discover unknown ones.

These next several months were a blur. Designing the war game series was a walk in the park. But their execution, and then making sense of the data we had collected, was challenging. We knew we needed to learn a great deal from our participants, and we mined the data generated by the wargames and symposiums very diligently.

The Chief of Naval Operations had selected the Naval War College because of its ability to hold multi-classification level wargames for large groups, and we were expecting well over 1,000 people. Over the years, this capability had been developed and regularly exercised in the Global War Game series held each summer since 1979. I participated in these games as a surveillance system and electronic warfare subject matter expert, and as a game director, from 1982 to 1988. I returned several more times in the next dozen years in my subject matter expert capacity.

As mentioned previously, for the games in the mid-1980s, I developed methodologies to input simulations of both space systems and special operations play as accurately as possible. Both of those capabilities have highly classified components; however, for this information to be useful on the main game floor, their effects had to be distilled down to the classification level of Secret, for which I also had developed methodologies.

The wargames of 2001 were a different challenge. Maritime infrastructure is both exceedingly vulnerable and highly significant to both the United States and the world's economic health and well-being. To defend such a far-flung target set, we needed input from many different maritime world elements, not just from the Navy and Coast Guard. The cross-discipline wargame was a device the Naval War College had pioneered. My officemate was responsible for the overall effort to respond to the president's task, and he asked me to lead the development of the game with the wargaming staff. I also acted as a subject matter expert on surveillance and communications systems, my core experience for over 30 years.

He also asked me to coordinate the dozen or so NWC personnel assigned to take notes, the rapporteurs. They are critical to any wargame success from which you want to extract lessons for follow-on actions. My job was to be sure they knew how to be a rapporteur before the games started and then taking their notes after the event and boiling them down to the lessons we needed to know and understand. Knowing and understanding are two entirely different things.

The first part of this task, designing the game and organizing the rapporteurs, was easy. The Naval War College personnel included many who had been rapporteurs before and were absolute pros at it. The second task, to refine the inputs into a new concept of operations for the nation, was the real challenge. As the lead rapporteur and the principal recorder, I worked with about a dozen others to capture the salient points and counterpoints generated during the games.

This last task is an essential function in any wargame, especially so for this one. From these notes, we would build the concept of operations to construct the president's answer. As the planning for the games progressed, I realized I would be right in the middle of something incredibly significant to the Navy, the nation, and the world. I was delighted to be doing something so worthwhile. On 9/11, we all felt useless. Now we had a focus. It was a great feeling.

It usually takes three to four months to plan and execute a significant wargame, but we had 12 days, and it was going to be a huge operation with many diverse elements. We all "fell to," as they say in the Royal Navy and were ready by September 24. It was a very hectic wargame because most of the participants had never played in one. But the wargaming staff had experienced this level of inexperience before and knew how to deal with it via training briefs at the outset and then coaching during the first day or so of gameplay. Thus, we were able to bring everyone into action by the afternoon of the first day.

The one thing that came out of the first wargame, which ran from September 24–26, 2001, was the explicit understanding that no agency tracked and identified ships until they were basically inside our harbors. We recognized that we needed to have more advanced warnings, and we generated a requirement to have all ships identify themselves 96 hours before entering port. The US Coast Guard very quickly published a Notice to Mariners establishing that rule. Interestingly, during those discussions, it appeared to me that I was the only one present who knew that the INMARSAT satellite communications system had a built-in position location reporting system that could be set to automatic.

The requirement to have ships identify themselves via that system when they were at least 96 hours out from arriving in the port and then keep broadcasting their position intermittently was the one solid recommendation generated by the first wargame. Taking the input from several people, I drafted a formal recommendation, and it was unanimously approved.

We decided to ask the International Maritime Organization (IMO), an agency of the United Nations tasked with responsibility for the safety and security of shipping, to implement this rule as soon as possible. The US Coast Guard took this requirement for action but warned that it might take some time as the IMO wasn't known for moving quickly. We all thought that the UN agency would move much more swiftly in these tense times than they subsequently did. The requirement was eventually enacted, but it took several years, not months. Its official name is Long Range Identification and Tracking, referred to as LRIT. Some see it as a competitor to S-AIS, but it is not; it is fundamentally different. LRIT requires a conscious action by the ship. It needs to consciously set the satcom system to report its position every four hours. S-AIS is designed to pick up the AIS beacon that all vessels over 300 tons engaged in international commerce are required to have on anytime they are approaching within 2,000 nautical miles of a coastline. Indeed, most ships leave it on as an extra automated lookout. AIS broadcasts its location and ship's identity at least every 10 seconds if the vessel is underway.

We held the second wargame on October 2–3, 2001. There were two primary results. At the end of the second wargame, I believed we had enough material to commence writing a draft concept of operations and told Dan Smyth so. He concurred and set me to drafting it. He was wholly tied up with the administrative requirements and was delighted I was willing to take this task. I believed it was the core task.

One of Dan's primary duties was to ensure the right people were attending and were assigned to the most beneficial roles. It is a lot harder than it sounds, what with egos, busy schedules of senior people, and all. You also have a limited amount of knowledge of the players' skills and personalities, especially in this series.

It was one of the most diverse groups of participants ever assembled, although the Global Wargames of the 1980s, run each summer there at Newport, were almost as varied. However, the Global games had the advantage of having most of its participants familiar with wargaming and the aspects of analysis. We did not. For this series, we had many civilians with no real idea of how to define the problem set, much less how to address it for analysis.

As we processed through the first two wargames, my team of several assistants and I tried to keep a combined, consistent record of the emerging main points. I immediately started organizing them into a rough concept of operations.

One fact stood out to me. We needed a better way to track all ships approaching our shores. During Game No. 2, Don Cundy, the electronic systems department head at the Coast Guard Research & Development Center, presented a brief on how they were going to improve their situational awareness of what was happening in and near our ports and harbors by using the Automatic Identification System (AIS).

AIS is a collision avoidance and shipping traffic control system mandated by the IMO required on all ships of the Safety of Life at Sea (SOLAS) class by February 2004. SOLAS class ships are defined as commercial ships displacing over 300 tons, all ships carrying six or more passengers, and all tugs over 600 shaft horsepower. Vessels displacing 300 tons were close to 100 feet in length, so most smaller vessels were not included, but individual nations have since modified downward the carriage requirements. As of 2018, the US carriage requirements are now down to all ships over 65 feet in length.

The Satellite-AIS aha! moment

As the Coast Guard R&D Center presented its brief in that second wargame, I thought back to my time as a space systems subspecialist in the Navy specializing in signals intercept and analysis. It instantly occurred to me that we might have the answer to our offshore tracking problem, and I asked Don Cundy if I could get an in-depth brief on AIS. He offered to give me one the next day at 0930 at his office in New London, Connecticut, about 60 miles from the NWC, and I readily accepted.

Early the following day, I drove down to Groton, arriving by 8:00 AM. Rather than ask to enter prematurely, I read in my car. From my two years as the officer in charge of the Naval Security Group Detachment, Atsugi, Japan, I knew that arriving early could be disruptive and might even be seen as rude. I knew Cundy had been with me in Newport for the previous three days and probably would like a bit of time in his office to catch up. I know I always did after a day or more away from my office.

They were ready for me when I rang the bell right at 0930, and Don had Dave Pietraszewski, his AIS expert, give me an in-depth brief and then a live demonstration of the signal. He displayed the AIS signal on a radar scope and had its data stream output to a printer. You could also click on the AIS icon on the radar scope, and all related data would appear on the screen, not unlike air contacts with their Identification Friend or Foe (IFF) system.

When the briefs were over, and they asked one last time if I had any questions, I asked the one question that had been running around in my head for the past 24 hours.

"Anyone ever thought about putting an AIS receiver on a satellite to give us a global ship identification and tracking capability?"

The answer from around the room full of engineers was that the thought had occurred and been dismissed. There would be too many signals of the same frequency and power arriving at the satellite's antenna simultaneously. All the satellite's receiver would hear would be white noise. The receiver would be totally overloaded.

From both my time at the National Security Agency and especially my time at E-Systems in Greenville, Texas, working on upgrading the communications system on Air Force One, I was sure I knew how the signal could be isolated and extracted from the noise. I also knew the guys who could write the algorithms to do it. However, I did not argue with the gentlemen in the room for two fundamental reasons. First, I was not totally sure I was right, and second, I was not sure of the classification of what I knew. It may well have been beyond Top Secret or, at the very least, company proprietary to E-Systems.

Besides, I was their guest, and that would have been not polite, so I just said I thought there might be a way around it and was going to investigate it a bit more. They skeptically wished me luck, and I departed for the 60-mile drive back to my office at the Naval War College.

Two other things pertain here: First, I had known since at least 1968 that ships did not have an identification beacon on them like aircraft with the Identification Friend or Foe system. And, about 11 years before, I had helped develop the plans for putting the first multi-band, phased-array communications antenna ever installed on an airplane—on Air Force One, at E-Systems in Greenville, Texas. I had also sat in on many discussions with its partner, Mitsubishi Electronics, in several places in Japan. Those discussions had included the subject of how that early phased array antenna would gain sufficient signal isolation from the signal noise floor. I thought that process just might be applicable here.

I hurried back to my office at the NWC and placed a call to Neil Cooper, one of the senior systems engineers at E-Systems, even before I sat down. E-Systems was the developer of the communications system on Air Force One from the very beginning, back when the aircraft was a Lockheed Super Constellation. Neil was as competent a man technically as I have ever met. I explained what I was thinking and what I now knew about AIS and asked his opinion. His answer was basically, "Physics is physics! Makes sense to me! Go for it!" So, I did.

I first used Google to look for an active satellite with a VHF system. There was only one, no matter how I asked the question: ORBCOMM, headquartered near Dulles Airport in northern Virginia. I immediately phoned ORBCOMM, but the first three or four people I talked to wanted me to sign a non-disclosure agreement before they would give me the time of day.

After those three or four unsuccessful attempts to talk with various people in their engineering department and executive branches, it dawned on me: I was talking to the wrong people. I phoned ORBCOMM back for the fifth time and asked to speak to their head of business development. The operator immediately put me through to Greg Flessate, director of business development. This time I did not start talking about technology. I started with, "Greg, I have an idea that could, conceivably, make ORBCOMM a lot of money! Are you interested?" Greg's response was just what I had hoped it would be. "You have my attention. Tell me more."

A Day's Collection from the ORBCOMM Constellation.
Courtesey of ORBCOMM

After about a 20-minute chat, he invited me in for a visit, as soon as possible, and I arrived there less than a week later. At that meeting, ORBCOMM agreed to place an AIS receiver as a "ride-along package" on their next satellite launch to replenish their M2M/IoT constellation—if I could find the money to design and build the receiver and cover the installation and integration costs. We were talking about roughly $10 million. The time window for getting this done was about 24 months. (Thankfully, over the next two years, that time window slipped about another two years.) Over the next 26 months, I briefed everyone I knew from my time with the Fleet Battle Experiments and from the several maritime terrorism wargames and seminars of late 2001/early 2002 asking for money to build my satellite, but all to no avail.

Just over a month later, in November 2001, the Coast Guard also asked me to help the Naval Undersea Warfare Center at Newport, part of Naval Sea Systems Command, install an AIS transceiver in a submarine preparing to depart for a special operation, and I did. We got them a specially modified transceiver with a three-position switch, On-Off-Receive only. The captain's post-mission report was all we hoped it would be: "Don't leave home without it!"

Chapter Eight

Networking to Success

For the next two years following 9/11, I briefed anyone and everyone I could find in the Navy and Coast Guard on both the utility of AIS and the need for Satellite AIS. I often mentioned the submarine test and got much feedback, a large part of it negative. Almost no one had ever heard of AIS, and the idea of putting a receiver in space sounded like craziness to most people. I distinctly remember saying many times that it would be an excellent idea to have AIS full-on, send and receive, in congested traffic areas where a warship's location and probable identification were already easily known. Indeed, I also distinctly remember explicitly mentioning the mouth of Tokyo Bay, which I had transited in two cruisers and twice in submarines, and repeatedly flown over in four different Navy and Air Force reconnaissance aircraft in my career. It was the most congested waterway I ever saw, but Navy friends tell me there are others, such as the Strait of Malacca, the English Channel, and Strait of Gibraltar, that are even worse.

Nearly the only encouragement I received was from my home organization, Johns Hopkins University's Applied Physics Laboratory. I was required to return once a month to discuss the day's issues and receive guidance from the Lab about what information they desired from me. My chain of command was Chris Latimer and Steve Biemer. Russ Gingras, the man who had hired me back in 1995, was my department head. I regularly conferred with them on the progress of my work at Newport and on finding funding for my wild idea. They all were encouraging, especially Russ, whom I had known since he and I worked on a project at the Naval War College about 16 years before. They all urged me to keep trying, as the prize was worth the effort. They referred me to specific Lab members who had experience with space technology, which allowed me to verify what I was considering was technologically feasible. I received excellent graphics art support from the graphics art department. The people there are true artists and knew how to sell a project with their skills. It was their work that I was dragging all over the US looking for money.

The other encouragement I received came from the Applied Physics Lab's technology research rival, the Naval Research Lab (NRL). I had briefed Pete Wilhelm, director of the Naval Center for Space Technology, with whom I had interacted during my time on the Naval War College staff back in the 1980s. He happened to attend a conference there about six weeks after I had the idea for S-AIS. I seized the opportunity and briefed him on it.

He liked the idea very much and put me in touch with Andy Fox, one of his principal assistants. Subsequently, we conferred often, especially on where I might find the money. The Naval Research Lab offered to assist ORBCOMM in designing their satellite, but that never came to pass, for reasons I never understood at all. Jealousy? Fear of compromising company privileged information? A combination of both of those? I never knew for sure.

By mid-November 2003, I had just about given up hope of ever finding money to build and install an AIS receiver in a satellite. Then I struck up a conversation with Capt. Tom Rice, recently retired Coast Guard, a new colleague in the newly created Maritime Domain Awareness Program Integration Office, an organization enroute to being a joint interagency, inter-departmental office hosted by the Coast Guard. Its goal: bring order out of the mishmash of maritime awareness programs across the government.

I had left Johns Hopkins, taking a significant cut in pay and benefits to come to work there a few weeks earlier, but this was one of the first times I had time to sit and talk one-on-one with Tom. He had recently retired as the head of Maritime Navigation Systems for the Coast Guard (and thus for the US).

As we sat discussing the task before us, he offhandedly remarked: "If we could just find a way to extend the range of AIS, our ship tracking task would become a lot easier." My response was entirely predictable if you knew my history for the past two-plus years.

This conversation ensued.

> "Are you trying to upset me?"
>
> "No, why would you ask that?"
>
> "Because I bet I have briefed at least 50 people in the USCG, including your deputy, about my idea of putting an AIS receiver on a low earth orbit (LEO) satellite which solves that exact problem."
>
> "Would that work? What about co-channel interference?"
>
> "I know people that are a lot smarter than me about this, and they assure me the signal can be isolated and read."
>
> "How come you have not told anyone about this?"
>
> "I bet I have briefed over a thousand people in the last 26 months, including every branch of the Navy and USCG, as well as several folks who worked for you."
>
> "No one told me anything. I would like to see your brief myself."
>
> "Don't have it here, but I have a CD sitting on my desk at home. I can bring it in tomorrow."
>
> "Great. See you at 0730 tomorrow. Right here."

The following day Tom Rice's immediate reaction to the brief was incredibly positive. His remark went right to the heart of the matter.

> "You need to brief the right guy."
>
> "OK, smart guy. Who is the right guy?"
>
> "Our new boss, Jeff High. The guy we were talking to yesterday."

"You obviously know Jeff High very well. How about setting up a time for me to brief him."

"Be right back," ... and he disappeared for about seven minutes.

Walking back into his office, his words were short and to the point:

"0745 tomorrow. Be ready!"

"I am ready now!"

"Jeff is an extremely busy man. He will see you tomorrow."

The following day, I briefed Jeff High right on schedule. He stopped me just less than halfway through my satellite AIS brief with that all-important question:

"How much would this cost?"

"I have discussed this in detail with ORBCOMM, and as a ride-along package with their planned constellation replenishment mission, it would cost less than $10 million to design, build, launch and operate the S-AIS package for five years."

"That is very reasonable if this satellite performs as you describe."

"Sir, that is what I have been telling folks for the past 26 months. It is a terrific deal, with an exceedingly high potential payoff."

"I am drawing a circle around $10 million in my budget right now. Get with Legal and Acquisition and come back with a firm number. Try not to spend it all! If you are right, this may well be the best $10 million the US Coast Guard has ever spent."

Breakthrough

After 26 months of searching, I finally had the break I needed. Dealing with lawyers and acquisition personnel is not my idea of fun, but they were very professional in assisting this newbie (me), and by the end of March 2004, we had a signed contract.

We took one additional intermediate step I believed to be very necessary. I asked for permission to fund a feasibility study at Johns Hopkins Advanced Physics Laboratory to ensure there were no unforeseen problems. (Jeff High believes it was his idea. Who knows; the outcome was the same.) I had already discussed this with Steve Biemer and Chris Latimer there and had a cost figure at the tip of my tongue for the Coast Guard: $110,000. I also requested permission to bring ORBCOMM into the study as they were the most experienced company in the country, and probably in the world, on VHF satellites. In that I had already written and gotten approved a Sole Source Justification for ORBCOMM, the Coast Guard Legal and Acquisition folks agreed, and we set up the team.

ORBCOMM sent their best, David Schoen, the principal engineer behind all three of the low earth orbit (LEO) communications satellite systems then in existence (Iridium, Global Star and ORBCOMM). He had rotated through employment at all three, developing the signal processing math for each satellite constellation in turn. He went on to be the Chief Technical Officer for INMARSAT, a major satellite communications company.

The ORBCOMM Constellation as of 2008. Courtesy of ORBCOMM

The Applied Physics Lab at Johns Hopkins reached into its deep pool of talent and assigned three young and brilliant PhDs. I do not think any of them were 35 yet. Maybe not even 30. We had our answer after less than six weeks of study, at least in part because I had spent the previous 29 months there at APL asking a number of very smart people there to look at the feasibility of S-AIS. Thus this study was just a confirmation. No one could find any reason why S-AIS would not work.

In order to save time and money I suggested we not write a formal paper. All Jeff High wanted was assurance we were not wasting $6,114,000 of USCG money. The Lab developed a formal presentation, and I briefed Jeff High with it. He signed the contract that I had worked out with the Coast Guard and ORBCOMM lawyers. I wrote over 90 percent of the contract, using examples from other similar contracts and then submitted my work for review by the Coast Guard lawyers and acquisition advisors. It is the only contract I have ever written. The rest is history, but even then, it was not all smooth sailing.

Just after we signed a contract with ORBCOMM to design, build, launch, and operate the first Satellite AIS system, the US Joint Spectrum Center published a technical paper highly critical of our APL paper. They said that S-AIS would never work because of the signal density.

It was quite a shock. I asked my team of bright young PhDs from Johns Hopkins to do a total review and comparison of our analysis—which said it was feasible—with theirs, which said it was not. I was very anxious that I might have just wasted $6.4 million. If so, my career in the science and technology field was over with a thunderclap.

I should not have been so concerned. I had a great team working on the problem. But it did take the team three days to find the difference in the two analyses. The Joint Spectrum Center had misplaced a decimal point in the definition of the signal density at that frequency. They had it 10 times denser than the acknowledged value. Our calculations were correct, and theirs were flawed. If you moved the misplaced decimal in their algorithm to the proper place, the two studies were in agreement. That night I got my first real sleep for the first time in three days. Still, this incident made me very anxious to see actual data from the satellite.

There was another existential threat to the S-AIS program. CG-2, Coast Guard Intelligence, had recently been formally admitted into the Intelligence Community and they had minimal background in sophisticated signals intelligence collection and analysis. Worst yet, they did not know what they did not know. I had the clearances to visit their spaces, and I did try to discuss how we could work together. But they did not want to hear any suggestions, no matter how gently I worded it.

There were a few retired Navy signals intelligence (SIGINT) professionals scattered about, and they all privately agreed with me. Still, the guys who had come up through the Coast Guard law enforcement pipeline were clueless about sophisticated data systems such as AIS, and all the ex-Navy SIGINT guys knew it. The CG-2 folks were dealing with people in this area at NSA for the first time. These were people that I had known for many years. Some of them had even worked for me or friends of mine and were now senior leaders there. They reported to me that the Coast Guard guys were daily demonstrating their lack of experience and knowledge. They were not stupid by any means, but they were ignorant. There is a big difference. The NSA folks pointed out to CG2 that they ought to work more closely with me, but that just made matters worse.

CG2 quickly became very jealous of my stature there and told NSA that I could not visit there without one of them being with me. The NSA guys just laughed out loud at that stupid edict and ignored it. I was in an inter-departmental organization, the Maritime Domain Awareness Program Integration Office. I ignored it too, but CG2 held my clearance, so it did get a bit sticky. Eventually, I asked Jeff High, my new boss and a Senior Executive Service Level 3, to help. In that he was senior to the head of CG2, who did not even know his juniors were restricting my access to NSA, that problem went away.

However, mid-level civilians at CG2, working with mid-level civilians at NSA, tried to kill the S-AIS program by getting it declared a SIGINT program—which would require a Sensitive Compartmented Information clearance to know anything about it. This, too, was laughable, but it became a direct threat.

The matter eventually blew up, and the issue went all the way to the Secretary of Defense. In late 2004, his office told both the NSA and the USCG (me) to prepare memos defending our positions. They could be no longer than three or four pages. I forget precisely how many because my paper was well less than the limit. NSA's took the limit. Mine said that I did not care what NSA did with the AIS signal behind closed doors, but there was a very valid need for it in the open, uncleared world. It was an Automated Aide to Navigation, a specific type of device that came under the purview of the USCG.

Also, at that point, I had been describing for about three years how I thought S-AIS could be used in the open market to anyone who would listen. And I had cleared my doing so with very senior people at NSA, in December 2001, before I ever opened my mouth in public. To try to classify it now was very much too late. I pointed out that the effort to classify it ex post facto alone would generate much more interest in AIS, and I did not think NSA wanted that. The cow was out of the barn. It is of no use to close the barn door now.

Friends tell me that a top DOD Under Secretary read both memos in a meeting at the Pentagon and immediately agreed with me. I breathed a big sigh of relief, and the plans for S-AIS went forward.

I understand from Canadian friends that when COM DEV, the parent company of exactEarth Ltd., the second company to operate a commercial S-AIS constellation, went to get permission from the Canadian government to create their S-AIS constellation, almost exactly the same thing happened in Canada. In both cases, common sense won out.

Russians fear 'CIA trick'

Today S-AIS is widely used across the maritime world and has, indeed, created the most significant paradigm shift there since radar, the steam engine, and the screw propeller. Yes, even more game-changing than GPS. GPS allows ships to navigate with much better accuracy, but it still left the ocean opaque. S-AIS is making the world's waterways much more transparent. It enables nations to understand who is approaching their coasts and ports, just as intended. Additionally, S-AIS has since become a ubiquitous tool for an ever-increasing array of maritime applications.

I had also been working with the Naval Research Laboratory on S-AIS since November 2001. I had known Pete Wilhelm, the Navy's Center for Space Technology director, for over 20 years. He was one of the first people I briefed. Pete was also the first person who completely understood what I was trying to do, and he was all for it. However, Pete worked in the classified world, and I wanted very much to keep this in the unclassified world, so we were on two different but parallel tracks.

It should be noted for history's sake that NRL beat ORBCOMM into orbit. They put an AIS receiver on TacSat-2, a proof-of-concept demonstration satellite they launched on December 16, 2006. However, it had a heat problem, and NRL had to shut down the AIS receiver after 20 minutes of operation on each orbit. But it did prove the AIS signal could be collected from space.

However, the mathematics Naval Research had embedded in the AIS signal processor were from classified spacecraft. So we still did not know if the ORBCOMM satellites would work or not. We thought they would, but we all wanted to see proof.

ORBCOMM originally planned to launch in late 2005, and it would have been the first S-AIS receiver in space. But there were several delays. ORBCOMM was finally ready for a mid-2007 launch of its six Plane A replacement satellites with S-AIS receivers via a Russian Cosmos rocket. However, at the last minute, Russia learned that some of the ORBCOMM-Coast Guard team had intelligence agency backgrounds (including me, specifically) and became convinced the ORBCOMM mission was some sort of CIA trick. We offered to give them full access to the downlink, but they were having none of it. They were sure we were spies. They backed the launcher off the pad, dropped it on its side, pulled the six ORBCOMM satellites off the bus, and put in bags of sand as ballast. They buttoned the launch vehicle back up, rolled it back onto the pad, erected it, and launched it without the ORBCOMM satellites. It took ORBCOMM about a year to finally get another ride—on June 19, 2008.

In the meantime, Space Quest, a small US company based in Fairfax, Virginia, was about ready to launch a test satellite with a programmable receiver in the VHF range, the same range as AIS, when they read a story about ORBCOMM, S-AIS, and the US Coast Guard. They decided to reprogram their receiver to collect AIS. They successfully launched the first S-AIS receiver—which could operate for an extended period into space—on April 17, 2007, six weeks after first hearing about S-AIS. Indeed, Space Quest purpose-built and launched two more S-AIS satellites, AprizeSat 3 and 4, on July 29, 2009. They are still in the satellite-making business today.

On 28 April 2008, Canada became the first foreign country to launch an AIS receiver into space with its Naval Tracking Ships (NTS/ CanX-2). Norway was the second foreign country to launch and operate an S-AIS receiver, collaborating with the Japanese to put their NORAIS receiver on the International Space Station which, after some problems with their antenna, became operational in June 2010, and launching their first AISSat-1 on 12 July 2010 about a month later. All these entities had attended our meeting on 22 April 2005 held by the multiagency Maritime Domain Awareness Program Integration Office at US Coast Guard Headquarters.

I consider that meeting the date S-AIS came of age and the concept of C-SIGMA was born. We called that meeting a Technical EXchange on AIS via Satellite (TEXAS I).

Game-changer takes the stage

There are now, in the summer of 2022, over 200 S-AIS receivers in space. The three primary collectors of S-AIS are ORBCOMM, exactEarth/Harris, and Spire. However, Spire has acquired exactEarth, so there are really just two major competitors in the S-AIS marketplace. However, the combination of these systems ensures that every AIS transmitter on Earth is monitored almost every minute. That time interval will fall to below a minute within the next few years.

S-AIS is now routinely paired with imaging space systems such as electro-optical satellites and especially synthetic aperture radar (SAR) satellites that can image the Earth day and night, rain or shine. There are not enough SARsats yet to do this 24/7, but with the increasing number of imaging satellites of all types, SAR, earth observation, video, plus the new radio frequency geolocation satellites (RFgeoSats), it is only a matter of a very few years before nearly every place on Earth, including the oceans and other waterways, will be under near-continuous observation/ surveillance from Space.

With all three types of these imaging satellites plus the RFgeoSats, (aka Electronic Intelligence Satellites or ELINT sats) there is a natural synergy in the maritime world as S-AIS gives the unique ability to identify the ships imaged or radar collected. Indeed, if they are not identifiable, the very lack of identity can generate curiosity and lead to further examination by other surveillance systems—from space, patrol aircraft, military, or law enforcement vessels. The same is true if they attempt to hide their identity via spoofing.

Indeed, attempting to spoof S-AIS is, in many cases, counter-productive because the spoofers are calling attention to themselves by the very act of sending out false information. In that many have the mistaken impression that S-AIS is easy to spoof undetected, a few words on that subject are necessary.

ORBCOMM could geo-locate all AIS emitters collected since their very first AIS systems were launched in 2008. That geo-location is compared with the reported position contained within the AIS transmission itself. If the two locations are more than a certain number of miles apart, the report is flagged, and that emitter is collected at every opportunity. Generally, once a position is verified to be "true," all following collection of that emitter by that satellite during that collection opportunity is discarded. (A collection opportunity for a transmitter is defined as when the sensor and its associated antenna on the satellite come into view of the transmitter until it leaves that field of view. Just to be precise, the antenna and the signal processing on the satellite defines the field of view.)

The data provided by the AIS signal includes the Maritime Mobile Service Identity. The MMSI is a unique nine-digit number that is assigned to each AIS unit. Some of these software tools contain databases that have archived the MMSI of every AIS signal detected anywhere, stretching back to early 2001. It is straightforward for these software tools to sort through the history of every MMSI ever broadcast in milliseconds and detect whether there is something amiss.

Those transmitters are also flagged for particular attention. Paired false locations and false MMSI generate extraordinary attention in law enforcement and intelligence organizations. There are also private firms tracking such activity for many different reasons. To be sure, not everyone has these tools, but they are readily available on the market. I would be delighted to introduce anyone interested in buying their services/tools to several developers. Most of them are friends of mine.

Another separate but related business area has also arisen. It is called by many names, but I prefer Static/Dynamic Data Analysis. S/DDA takes the reports from many sources, satellite and terrestrial AIS, as well as from the myriad static information and data sources in the maritime world. Those sources include ship movement reports, brokers' reports, financial news and shipping documents, as were S/DDA strives to make sense, understand, and (hopefully) gain wisdom from this mass of data. AIS is its core data source. Many of these companies now use artificial intelligence and machine learning to develop their reports. Some of these systems such as Space Eyes, can provide information to their customers in a tactically useful timeframe.

These software tools process S-AIS data for many uses. Radar and satellite navigation—even in its most recent form, GPS— are the only other systems that even come close to the impact of S-AIS in the maritime world in the last 150 years. However, while GPS allowed mariners to navigate with more surety, it has left the maritime world opaque even as S-AIS quickly makes it much more transparent.

Like satellite navigation, which the Johns Hopkins Applied Physics Laboratory created to improve the submarine-launched ballistic missiles accuracy, S-AIS is rapidly becoming ever more present in the marine world as more and more applications for its data are developed. However, while GPS is now embedded in many applications for all three environments—sea, air, and land—S-AIS's impact is focused on the marine world, bringing significant changes in its global operations.

Indeed, as far as the maritime world's overall operations, the impact of S-AIS in its first dozen years of existence may be more significant than satellite navigation at this point in its evolution and has most probably already surpassed it in total, judging by the breadth of changes.

Commodities. The tracking of all the world's commodities location and the estimate of their time of arrival at the destination is now down to a very few hours. This detailed information, derived from S-AIS and brokers' records, has allowed commodities traders to be dramatically more accurate in predicting daily prices in many ports of the world, thereby allowing the first few who developed and employed the analysis based on this information to reap huge financial rewards. Now that everyone is doing this, the field is much more level.

Marine Maintenance. S-AIS helps determine when hull and machinery maintenance are needed. By accurately reporting a ship's speed, it allows for accurate record-keeping of hours of operation at what loading/speed and in what type of marine environment at what average speed (critical to hull maintenance). These calculations are saving the big fleet operators millions of dollars, and reducing their carbon footprint.

Illegal Fishing. Satellite identification shows when a ship is bound for an area with environmental or resource sensitivities or might traverse an area known to be contaminated with a sea life-threatening biological problem, such as a disease or predatory fish. The patterns of the courses used by fishing boats while fishing are very distinct in many instances. Several software tools now automatically recognize these patterns. Global Fishing Watch, the Pew Foundation, CLS, (a subsidiary of the French Space Agency), e-GEOS, a partner of the Italian Space Agency, and others have built an array of software tools to aid this effort.

Environmental Protection. Using synthetic aperture radar satellites to detect illegal bilge dumping in controlled waters is now much more effective because the exact identification of the offending vessel and its next port of call can often be determined via S-AIS. When the vessel docks, local authorities can demand to see both its bilges and its log. If the bilges are clean, but there is no log entry as to when they were last pumped clean then a citation, often worth many thousands of dollars, is issued. The Italian Navy reports this has caused a dramatic reduction in the illegal dumping of bilge waste, thereby improving the marine environment in the Mediterranean.

Search and Rescue. S-AIS has also dramatically improved safety of life at sea by allowing for the location of all ships in an area to be known by all interested parties thereby permitting the rapid reaction to maritime disasters, large and small. Indeed, AIS transmitters are now being installed on life jackets to assist in the recovery of crew in the water. S-AIS allows for the closest vessels to a maritime problem to be identified and vectored to the site as needed.

Disaster Mitigation and Recovery Support. It is also a significant tool in disaster recovery operations. In Hatti, S-AIS provided the knowledge of when needed supplies would arrive at ports which were severely damaged and thus had minimal offload capabilities. S-AIS allowed for the accurate planning of the landing of the most needed supplies in priority order. It has been used similarly in many disasters since, including the Philippines, which has been hit several times recently with hurricanes, and in the Indian Ocean, Japan, and Chile, where tsunamis have damaged ports and seacoasts. It is now standard operating procedure in these situations.

Security and Surveillance. Finally, S-AIS is being used worldwide as a primary adjunct to maritime security operations by allowing for the study of regular pattern-of-life operations to determine when anomalous activity is happening. Surveillance and intercept units can be dispatched with much-improved chances of apprehending wrongdoers, be they smugglers of all types, or illegal fishers, or perhaps illegal immigration. This saves wear and tear on equipment and personnel, as well as money, by limiting the operating time of scarce assets. It also raises crew morale because it knows they have a better chance for a productive operation.

Sanctions compliance. The accurate pattern of life at sea that S-AIS allows to be generated is now being used to identify violators of international sanctions. Here too, AI and machine learning is being employed with good use. And more. The list just keeps getting longer. But clearly, S-AIS is making a significant impact on the maritime world. I must admit, I am enormously proud.

S-AIS and machine-to-machine (M2M) satellite communications, have recently been combined into a revolutionary package the size of a square beer can. The device is called the MT5000. It was originally known as Hali. With an 18-inch whip antenna, it was designed for the small-sized fishing and pleasure boats of the world. It transmits low-powered AIS to alert larger ships and shore stations within 10 to 12 miles as to its location as a safety device. It also broadcasts its position to ORBCOMM's 40 satellite M2M constellation at the same time to allow for the relay of this data to a fleet operator or family members, or anyone else the mariner wishes to keep informed of their location. The uses for this new device, the MT5000, have just started to be explored, but I believe it will be another game changer in the small boat world.

NOTE: Both satellite navigation and Satellite-AIS were instigated at Johns Hopkins University's Applied Physics Lab (JHU/APL). I was an employee there at the time I conceived AIS and the encouragement and the assistance that I received from there in many ways as I struggled to sell my idea, were invaluable. I might well have quit looking for funds had I not had such solid backing from Russ Gingras in my home office, and from Pete Wilhelm, director of the Naval Center for Space Technology.

Chapter Nine

A Call for Global Cooperation

Why C-SIGMA—Collaboration in Space for International Global Maritime Awareness— matters.

Illustration of my concept that would serve as a guiding principle for maritime security. Courtesy of author

The idea for C-SIGMA grew directly out of satellite AIS. Once the Maritime Domain Awareness Program Integration Office made the announcement that it had funded ORBCOMM to build the first satellite AIS systems, we very quickly gained the attention of others, including the governments of Canada, Norway, France, Germany and Japan. We also had MDA of Canada, KSAT of Norway, Airbus of Europe, DLR of Germany, e-GEOS of Italy, and JAXA and Mitsubishi of Japan contact us to discuss working with them to build a space-based maritime domain-situation awareness system.

With the exception of Norway, each organization either had, was building, or both, a synthetic radar satellite. And each wanted to build their own maritime situational awareness system. You did not need to be a rocket scientist to immediately see the advantages of a global coalition collaborating to build one great system rather than many simply good ones; consequently, the C-SIGMA concept was born. I gave a presentation on C-SIGMA in the spring of 2006 at a conference on space and national security in northern Virginia. It was very well received, and I have continued to update and brief on the advantages of unclassified space-based maritime awareness ever since.

Canadians were the very first to come to see me about teaming to build a system. Both Lieutenant Commander Robert Quinn and Dr Jake Tunaley, members of the Canadian Forces Space Command, and Andre Stavri, a representative of MacDonald, Dettwiler & Associates, Canada's leading space systems builder and operator, visited me several times starting in mid-2004, both separately and together, to describe their RADARSAT-2 satellite being built near Toronto and the Polar Epsilon program, their program for maritime awareness. Several other employees of MacDonald, Dettwiler and Associates also visited me several times that year, but I mainly dealt with Andre Stavri.

Both the Canadian government and MacDonald Dettwiler were interested in putting the AIS receiver ORBCOMM was developing for us onboard RADARSAT-2. However, after much discussion over many months, it was decided that while they could accommodate the AIS receiver, it was too late to modify the satellite to accommodate the VHF antenna needed to receive the AIS signal. I suggested they put a small satellite up in the same orbit to fly in loose formation (as the Soviets did with their RORSAT/EORSAT duo) and that was the start of exactEarth, the initial major competitor to ORBCOMM.

We all stayed in close touch and in the late spring of 2007 Andre Stavri called to ask if I was aware that the US Air Force had leased RADARSAT-2 for three months during its systems checkout just to run tests and evaluation of its utility for various military missions. It was to be launched in December 2007 and the planning how to conduct the tests was going to take place at Offutt Air Force Base south of Omaha in about a month. The Air Force and the US Army were the only ones who had been invited to that planning meeting.

Andre wanted to know if I would be interested in attending. I jumped at the chance and he pulled some strings, and I was invited but only as an observer, with no active role. I went to the meeting and sat in the back listening to a day and a half of briefings on what the Air Force and Army were planning to do to test this new radar satellite. Andre was entirely right, there was no plan to test RADARSAT-2 anywhere near water. At the end of the conference, I went up to the colonel running the test and evaluation of the RadarSat, identified myself as the US Science & Technology Advisor for Maritime Domain Awareness and asked why this was so.

"I have no one with any experience in maritime operations to write a test plan," was the answer.

I asked him if he knew what the RC-135V/W Rivet Joint was.

"Of course! We have a squadron of them parked at the Air Force base here."

"Would you happen to know who wrote and executed the Initial Operational Test & Evaluation of that airplane, by any chance."

"No, I am sorry I do not."

"You are looking at him, and I would be happy to do the same for you," was my reply.

I suddenly had his full attention. "Nice idea, but we have no maritime targets to image. The Air Force and the Army have no boats."

"Colonel, I have about nine years at the Naval War College over a 25-year period. I was a student there and used to lecture and do research there, a lot of it related to space operations. And my office is in Coast Guard headquarters. I bet I can find you boats."

"If you can, I will give you at least a quarter, maybe a third of the time on the satellite. That would be a particularly useful use of the RadarSat's time."

And that is how RADARSAT-2 came to be involved in overwater operations.

Satellite the size of a bus and a Florida drug-runner

It was quite easy. I contacted the Navy TENCAP (Tactical Exploitation of National Capabilities), the Office of Naval Intelligence, and the numbered fleets (2nd, 3rd, 5th, 6th, and 7th), all of which I knew from my previous assignment to the Naval War College's research side, and innocently asked: "Anyone there want to test an unclassified synthetic aperture radar satellite over water?" and got out of the way. Otherwise, I would have been trampled by the mob.

I also asked the Joint Interagency Task Forces South and West if they were interested. JIATF South was most interested, but they were completely out of discretionary funds to buy fuel for their test boat. My office was especially interested in testing against their test boat as it was a captured 30-foot-long drug runner. It would be the hardest target we would have for the whole test series. In the end we gave them the funds to buy the fuel for their boat out of our office supplies funds.

The Navy TENCAP office, where I had worked on the IBS program in the late 1990's, coordinated the tests with the numbered fleets' offices and the TENCAP project office, and I went to one of the final planning meetings in Canada. We even got to go into the final assembly room and stand next to the fully ready satellite. It is about the size of a 45-passenger bus. We could have reached out and touched it, but they told us if we did, we would leave our hands behind as they would cut them off. I think they were just kidding. (Maybe not.)

The one test I ran in its entirety was the test of RADARSAT-2 against the drug runner off Florida. We ran the series of live tests in late February and early March of 2008, three-plus months after the satellite went into orbit. It was completely successful and clearly proved that a SARSat could detect at least the engine—as well as the wake—of a typical smuggler boat. MacDonald Dettwiler was delighted.

The rest of the tests in the various numbered fleet areas of operations were also successful. It was tested off San Diego, Hawaii, Japan, in the Mediterranean and in the Gulf of Guinea. I correlated the test reports and forwarded them to the Air Force colonel running the whole show. He was asked to brief both the Chief of Staff and Secretary of the Air Force, the two most senior Air Force people, and their senior staff, on the results of RADARSAT-2 test. The colonel asked me to review his brief and then to be present in the anteroom of the briefing room both to answer any last questions he might think of and to be on call to respond to any questions the two leaders might have.

I felt particularly good to have gone from an unknown guy in the back of the room to basically the front row, but it was the colonel's remark as he left me to go in to brief the two USAF leaders that really made me feel great.

"Hate to admit it, Guy, but you Navy guys did a better job of wringing out this satellite than anyone else, and I am going to say just that." Then he walked in to tell a room full of very senior Air Force brass the same thing.

'Like Mick Jagger, Brad Pitt'

Every even-numbered year MacDonald Dettwiler holds a meeting at its world headquarters near Vancouver, British Columbia. They call it the Earth Observation Business Network and it is by-invitation-only. But MDA invites every organization in the world which owns or operates an Earth observation space system, including Russia and the Peoples Republic of China as well as the Republic of China, to attend.

In May 2008 I was honored to be invited to participate and speak on our recent tests. To put the tests in context, I included them at the end of the C-SIGMA core brief, which deals with how commercial space systems could be used to form a global international collaboration for the good of the world. Most of the participants were companies and organizations focused on the land masses of Earth. But I thought I would try to get their attention on the utility of space systems for maritime situational awareness possibilities now available with the advent of S-AIS.

The Vancouver meeting was scheduled to start early on a Monday, and as I was getting things in order to leave on Friday afternoon, I received a phone call from a senior Vice President of MDA. He had just read my brief, which I had sent ahead as requested, and really liked it. Would I mind if my speaking slot was moved to be the first speaker of the whole conference (i.e., the keynote). He thought the ideas I presented just might hit the right innovative note to kick-off the conference. I was delighted to accept, of course.

As requested, on Monday morning I gave my talk right after the welcome from MDA's CEO. For the next three days I was treated to the kind of attention Mick Jagger and Brad Pitt are used to, but it was heady stuff for an ex-deck force sailor from Corpus Christi, Texas. Many of the space systems operators immediately saw the potential revenues to be generated if they could operate their satellites over water. Space-based maritime situation awareness was born in a flash there in Vancouver, British Columbia on the first Monday in May 2008. Suddenly Germans, Italians, French, Brazilians, Chinese and Japanese (especially the Japanese) were my best friends.

As related in a previous chapter, in August 2009 I was asked to be part of the team writing the National Space Policy, Presidential Policy Directive Four, of the Obama administration. After nearly nine months of effort, during which my work was singled out repeatedly as being among the most original, innovative, and valuable of anyone's, wording was inserted in the draft ordering the implementation of basically the C-SIGMA concept.

It was fascinating to me that my strongest supporters were the military and the Intelligence Community members of the team. The Department of Transportation and the Federal Aviation Administration were also significant supporters, and the Department of Commerce was not far behind. I think intelligence experts saw that C-SIGMA would lessen their load and help satisfy some of the taskings they were getting from the Navy. Transportation knew there could be considerable benefits to be gained if the location and status of all vessels at sea on a global scale could be determined in a sharable data manner.

In the Old Executive Office Building, a bugaboo is back

The final review of the National Space Policy was held at the Old Executive Office Building on the grounds of the White House on June 21, 2010. We were all surprised to learn that all implementation directives, classified and unclassified, were to be bundled into a classified implementation directive. Back at the beginning of policy formulation, in August 2009, three core points were clearly spelled out in the tasking letter. The three points were, to paraphrase:

—Use Space as a means to bring all nations together for the good of all mankind. (This was what C-SIGMA had set out to do since the beginning back in 2005.) The State Department had the lead, and this was the team I had been part of from Day One

—Be sure that we maintain a robust space industry. The Department of Commerce had the lead.

—Support our national technical means of treaty verification for arms control treaty compliance and ensure we have the absolute best systems available. The Pentagon and the Intelligence Community were co-leads.

There were 15 implementation tasks about evenly divided among these three salient points. Only five or six were classified, and I was utterly puzzled and dismayed.

I raised my hand to ask the obvious question. Why was this being done? It would compromise all the work of the group working on the first point of the tasking. I went on to say that the whole point of the concept of C-SIGMA, and everything else done by the State Department group, was for the open and free exchange of information between all participants. If the tasking is classified, it would be nearly impossible to implement any of the things we have been discussing for the last nine months.

The answer blew me away. The White House did not want either Congress or the press asking them questions about the implementation tasks and criticizing them if some of the implementation tasks fell behind schedule.

Then I really stepped in it. I was so angry that I asked a second question when I should have just shut up and sat down.

"Whose stupid idea was this?"

People tell you never to open your mouth when you are mad, but I was too angry to remember that sage advice. The answer was even more startling than the first:

"It was the president himself! You will sit down, Mr. Thomas!"

Less than 18 months earlier President Obama had campaigned on having the most open administration, and it was all a lie. I was invited back to an implementation meeting only once more, to the celebration of the signing. Other than that, my time with the White House was at an end.

Implementation Task #1 of the National Space Policy of June 28, 2010 tasked the National Maritime Domain Awareness Coordinating Committee to build the envisioned C-SIGMA system. Unfortunately, that committee was chaired by someone who, as I discussed earlier in this book, has uncalled-for personal animosity toward me. He believes, erroneously, that I stole his ideas for the National Strategy for Maritime Security.

Because of his animosity towards me he has ignored the direction of the president as contained in the implementing directive and pigeon-holed the task. It has never resurfaced. A civil servant with the equivalent rank of a one-star general has thwarted the direction of the president and gotten away with it. I am still amazed, but I am told it happens all the time in a massive bureaucracy such as our government. But it still hurts when it is your plan that is killed that way.

We are still actively looking for a way to move the system to implementation. We have briefed the concept many times on every continent except Antarctica, as well as to the European Parliament. We were asked to attend Ocean Week 2017 at the United Nations, to discuss this, but, unfortunately, the letter confirming the invitation was misrouted and did not arrive until after the event.

Chapter Ten

The Case for
a Global System

*The Final Step: Fruits of collaboration
and wisdom of C-SIGMA*

No single country can do it

Maritime Security has different dimensions, including but not limited to Maritime Situational Awareness (MSA): law enforcement; maritime safety; maritime environment; marine science and technology; maritime trade and economy; maritime law; and public health. Therefore, in national terms, maritime security can only be achieved by a whole of government approach. If we succeed in applying this approach together with like-minded countries in a multinational environment, the C-SIGMA concept maintains that we can attain our common maritime security objectives.

To do this, C-SIGMA envisions linking together existing and planned unclassified space system capabilities in a worldwide collaborative network via coordinated regional centers for international global maritime awareness to provide security, safety, environmental protection and resource conservation, as well as disaster mitigation and recovery. It would be a huge step toward global maritime security.

In the 21st century it is well known that the cyber world has expanded exponentially. However, unnoticed by many, since 2004, and increasing steadily since then, there has also been an ongoing revolution in space-based Earth observation systems. Led by space-based AIS, their utility over the world's waterways has increased dramatically. These capabilities not only support safety and security at sea but can also significantly assist in economic and environmental stewardship and resource protection, as well as disaster mitigation and recovery. This is especially true of the remote areas of the world such as the Arctic, and the resource-rich regions of the underdeveloped world such as the Gulf of Guinea, the South China Sea, Micronesia and the Indian Ocean.

The potential contributions of space-based Earth observation systems to maritime awareness are of growing interest to the world's naval and law enforcement forces, as well as to environmental preservationists. Also: governmental transport; commerce; maritime; environmental protection; and disaster preparedness ministries, in addition to shipbrokers; and others with an interest in the marine domain, its environment, and the protection of its resources. But coordination to maximize these capabilities is lacking.

Ongoing research that started just after 9/11 shows that no single country or international organization has the ability and resources to fully support the safe, secure, and efficient use of the maritime domain as well as the conservation and protection of the marine environment with its finite resources of fish, minerals, and oil, as well as to substantially assist oceanic commerce.

In that no one country has sufficient resources, including spacecraft, strong international collaboration is essential to achieving these objectives in a balanced manner. Indeed, this effort may need to be managed by an agency of the United Nations. However, agencies of the United Nations have shown themselves to be very cumbersome and slow to react to changing technology. Earth observation satellite capabilities as well as data information handling capabilities are expanding very rapidly, with no end in the foreseeable future. So the UN might not be the correct organization to take on such a dynamic task.

There is great need—and significant opportunity for international collaboration—in national and regional efforts to coordinate the use of the space technologies that are now available for detecting, identifying, and tracking vessels. These systems are especially suited in areas with shared international interests, such as the Arctic, or in pirate-infested waters, or in areas known to support smuggling or resource theft.

By 2022, there was a virtual tidal wave of new space-systems with improved earth observations capabilities being built and launched, with even better satellites being planned. Additionally, as indicated above, the cyber world is also enjoying a similar expansion. The time to take advantage of these significant opportunities rapidly bearing down on the world and on the horizon is now, as this tide of technology rises. Catch the wave! Seize the moment!

How it works

There are seven elements that must be integrated for effective results. (Five in space, two on the ground.) Two of the five different satellites types employ imaging sensors:

1. Synthetic aperture radar satellites (SARsats)
2. Electro-optical imaging, both still and video, satellites

The other three satellite types are systems based on the radio frequency spectrum:

3. Individual transponders sending short, formatted status reports to communications satellites (M2M).
4. Automatic Identification System (AIS), an automated short message system designed for collision avoidance and traffic control in congested waters now also is used globally as a primary ship identification and tracking system. It is the crucial component of this concept.
5. Radio Frequency Geolocation Satellites (RFgeoSats aka ELINT sats) that detect and locate both the radars and communication systems of ships.

The other two elements are more prosaic:

6. The ground infrastructure: terminals, software tools and licenses to allow users of the system to determine which spacecraft to task when to obtain the desired results, and to act.
7. The software tools to correlate, fuse and analyze the information generated by the space systems, including

S-AIS track data, the basis for all analysis, along with other pertinent data resident in all reachable data sources such as port, financial, shipping and broker records.

While C-SIGMA would go a long way toward satisfying many of the world's varied needs for maritime situational awareness, it would have the added benefit of providing a focal point for the creation of the global maritime security system envisioned in *A Cooperative Strategy for 21st Century Seapower*, a policy statement of the US military maritime services.

The coordination needed to implement C-SIGMA would provide a focus for the efforts to achieve a common goal of protecting the maritime environment, as stated in that document, and would go far in bringing the lawlessness of such places as the Gulf of Guinea under control. Space-based earth observation does not replace terrestrial systems but does make them substantially more effective, and it is a significant start toward fulfilling the core need of knowing who is where on the world's waterways.

C-SIGMA also directly supports the US National Space Policy of June 2010 and could be the international mechanism to satisfy its Implementation Task Number One. That policy emphasizes United States leadership in space and directs international collaboration on mutually beneficial space activities for the purpose of broadening and extending the benefits of space to all mankind. Indeed, Task Directive Number One orders the Secretaries of Defense, Homeland Security, Transportation, State and Commerce to develop an unclassified program to foster international collaboration using unclassified space systems to enhance safety of life at sea; increase mutual security of all users of the maritime domain; improve protection of the maritime environment and the resources of the sea; and improve flow of commerce. It also mentions using the envisioned system to better monitor the Marine Transportation System, which is not pertinent here.

Why the National Security Council has ordered the deletion of this unclassified paragraph is a complete mystery to me.

While space-based MSA is indeed catching on globally the implementation of that directive has been held in abeyance for some unknown reason. However, implementing C-SIGMA could well be the key to building the envisioned, genuinely global, maritime security system. It would substantially assist in many tasks, including detection of illegal smuggling of all types of goods, arms, drugs, and people; much improved maritime pollution control and resource protection, such as illegal fishing and stealing of oil; dramatically assisting humanitarian assistance and disaster recovery operations. Remote ocean surveillance in such areas as the Arctic, the southern oceans, and the western Pacific would clearly immediately benefit many people both in and out of those regions.

Implementing C-SIGMA in the Arctic and a few other locales such as the Gulf of Guinea and the western Pacific could be the needed steppingstones to the implementation of Global Maritime Awareness for the betterment of the entire world.

In Chile, fruits of collaboration

One of the first tests of the concept took place in Chile. In 2008, Chile hosted the Western Hemisphere Maritime Domain Awareness Conference and a year later, the Global MDA conference. C-SIGMA and Space-based Maritime Situational Awareness were discussed in detail at both events, and Chile later asked for a briefing for their Naval War College faculty and staff. Chile offered to conduct a live, limited objective experiment to test the concept (if they did not have to pay for the satellites' input).

In 2010 the United States Office of Global Maritime Situational Awareness, in conjunction with the Office of Naval Research Global, set up a three-month test, with ORBCOMM also a major player. During the first month of the test, three noteworthy events took place which we believe proved the case.

Chilean fishermen reported numerous foreign fishermen in Chilean waters off Easter Island, 1,500 NM off the South American coast. The S-AIS and SAR systems were tasked to search that area and reported no unusual number of vessels in that area. The test director thought the test had failed until the Chilean admiral in charge of that area thanked him profusely, explaining the information has saved him the cost of flying a C-130 maritime patrol aircraft to search the area and sending a warship to investigate.

At about the same time, a Chilean open-ocean fishing ship had suffered significant storm damage several hundred miles offshore. The vessel lost its satellite communications and navigation systems and the only electronics still operating were its high-frequency communications and automatic identification systems. Worse, one of its crewmen had a compound fracture of his leg and needed immediate attention. But the ship did not know its own location. ORBCOMM was able to provide that information from its AIS, and a destroyer with a helicopter was able to reach the ship and rescue the crewman, probably saving his life.

Then S-AIS reported, and synthetic aperture radar confirmed, that a large fleet of Chinese fishing vessels was approaching the Chilean coast. The Chilean Navy prepared to intercept, but the foreign ministry asked for a delay to see if something could be done. They discussed the issue with the Chinese ambassador, relating they had firm proof from space systems that the fishing boats were in Chilean waters. Chile was prepared to impound the illegal fishermen and their vessels, the Chilean diplomats said. The Chinese ambassador denied all knowledge of the vessels. But soon thereafter, the Chinese fleet turned north, into Peruvian waters. By the time Peru could react, the illegal fishermen had moved on.

The overall test was a complete success with the Chilean Navy and Coast Guard becoming firm believers of the utility of Space-Based Maritime Situational Awareness, which is another way to say Geospatial Information in the maritime domain. Peru too is now a firm believer and Brazil and Argentina have followed suit, but the real task is to get all nations of South America to form an alliance to use Space-Based Maritime Situational Awareness to jointly protect their marine environment and its resources, as well as to provide increased safety, and security from smugglers of all types. That is already being done in Europe with its European Maritime Safety Agency and Frontex, by which the European Union controls its borders.

Maybe the United States will eventually join the club as well.

A Russian liner adrift

In early 2013 we got the chance to again demonstrate the utility of the collaboration of several countries' space assets in maritime awareness operations. The MV *Lyubov Orlova*, a 285-foot Russian Arctic liner named for a Russian movie star, had broken down in St. Johns, Newfoundland, in 2010 and two years later was sold for scrap. The new owner learned that the Canadian government was considering impounding the vessel for non-payment of bills and decided to tow it to the breaking yard in the Caribbean in January 2013.

Most experienced people understand towing a ship in the North Atlantic in the winter is not a good idea. Folks in the maritime business also understand you want ballast in your ship at all times, especially when it is not under power—when it is being towed, for instance. But, for whatever reason, the owner decided to get underway in January, and he had both the fuel and ballast tanks emptied and sealed. Not long after the ship left St. Johns, quite predictably, it began hobby-horsing, swiftly moving up and down, adding significantly to the strain on the tow cable. Quite predictably, the tow broke. The seas were rough; the tug could not regain its tow and returned to port.

A second, larger tug was contracted to regain the tow, but it, too, failed. In the meantime, the Canadian Coast Guard learned that the MV *Lyubov Orlova* appeared to be drifting into the area of a large offshore oil-handling facility. Everyone was concerned. Transport Canada, the ministry that operates the Canadian Coast Guard, decided to dispatch an even larger tug to regain control of the vessel.

The cruise ship also broke that tow, but not until the latest rescue tug had moved the derelict vessel far enough away so that the oil facility was no longer in danger. The tug returned to port and the Canadians maintained an aerial watch on the ship for a few more days as it drifted away on a northeasterly course. When it was determined that the vessel was no longer a threat to Canada, surveillance ceased. A dozen or so days later, Canada decided it had a duty to inform other countries in the North Atlantic that there was a large ship adrift that appeared to be headed their way.

I had been asked to put together a panel on Space-Based Maritime Situational Awareness at a meeting at the European Parliament in Brussels on the same day Canada decided to inform its neighbors. Rear Admiral Chris Reynolds, the director of the Irish Coast Guard and an early C-SIGMA supporter, was one of my panel members. As the meeting started, he turned to me and said he was going to have to leave immediately after he spoke.

He had just been informed of the *Lyubov Orlova* situation and he needed to get home ASAP. At that time, the best estimates were that it would hit the northwest coast of Ireland.

I replied that this looked like a golden opportunity to prove the utility of the C-SIGMA concept. He liked the idea but had no budget to support it. I made a deal with him then and there: If I could get ORBCOMM and the owners of the three active SAR satellites to donate their services in the name of international maritime safety, and I acted as the overall coordinator, would he assume overall command?

"Done," was his reply.

It was an easy sell to all four companies, but the German Space Agency had no time available on its satellites and MacDonald Dettwiler was tied up for several weeks. But ORBCOMM and e-GEOS, the Italian company, leapt to action. We asked Transport Canada for their draft models, and then the US Coast Guard and the National Oceanic and Atmospheric Administration, as well as Ireland's and the UK's Coast Guards, signed on to assist us.

My next month plus was very busy. I have a file of several hundreds of emails, and my phone bill was outrageous. The problem was that no one had drift models for a ship this large. It had never occurred to anyone that a nearly 300-foot-long vessel would go missing for anything like four weeks, and that was the time interval we were dealing with. The US Coast Guard made some changes to their model, and we started coordinating with e-GEOS and ORBCOMM.

For the first 10 days or so ORBCOMM was able to correlate every satellite detection to an AIS detection. Then we received the coordinates of one of the Orlova's Emergency Position Indicating Radio Beacon (EPIRB), and a few days later a second. We found nothing at one location but received an image from e-GEOS of what may well have been a lifeboat at the other. Still no vessel, but we were getting warm, maybe. I now know that objects of different sizes drift at different rates.

Shortly after, the owner called to assure me he was very sure the vessel was still afloat. That was when I learned that the owner had emptied and sealed the fuel and ballast tanks. My opinion was just the opposite of his. All vessels need to have ballast to stay upright in heavy seas. Indeed, ships will pump seawater into their fuel tanks in heavy seas to gain and maintain stability. By emptying both the ballast and fuel tanks, the ship was clearly unstable, and it was highly likely it had rolled over and probably sank some weeks earlier. The fact that two EPIRBs, one floating by itself, and the other probably in a lifeboat, had been located at widely separated areas strengthened this conjecture. I quietly called off the effort. We were wasting our time.

About a week later the Irish Coast Guard got a report of a sighting of a large smooth object about the size of the upturned hull of the *Orlova* and we asked e-GEOS to resume its radar satellite search in that area, well northeast of where we had been looking. ORBCOMM also resumed sending all involved reports of radar detections not seen broadcasting AIS. We didn't find the Russian vessel, but we did image two fishing boat-sized contacts with abnormal tracks and not transmitting AIS. The Irish Coast Guard decided they might be illegal fishers and launched maritime patrol aircraft to investigate. One evaded detection, probably by resuming a normal track, but the other turned out to be an illegal fisher from Spain and was apprehended. So all was not lost.

Indeed, it was a good experience and training exercise. Transport Canada was especially appreciative of our efforts, and the Irish felt they had learned a good bit, too. It must have been a success as I subsequently learned that others were claiming credit for the effort. And none of them had anything to do with it!

Indeed, at a conference in Japan in 2017, I heard someone else, who I am quite sure had absolutely no role in the effort, claim credit for coordinating the space systems search.

Sigh!

Chapter 11

Eyes Above: Operations in the Global Commons of Space

Where we must go from here

Maritime security has been essential all over the world for the proper functioning of maritime domain since man first went to sea as fishers, and then as traders. With the high degree of inter-connectivity now existing throughout the modern world, security in all domains has become critical at all levels and for all roles, and all security begins with situational awareness. Nowhere is this truer than in the oceans, where correct functioning is critical to the well-being of the world.

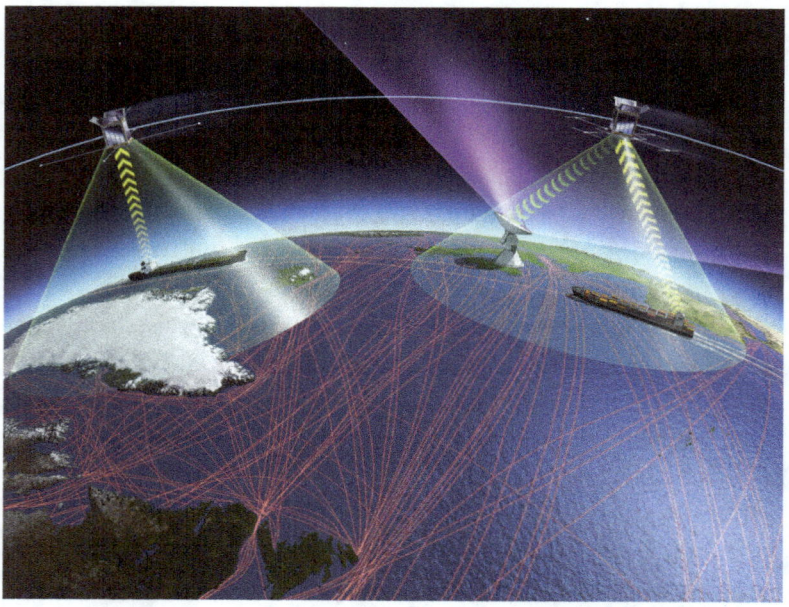

SAT AIS Satellites, Artist Impression. Courtesy of ESA – European Space Agency

Where three global commons converge
for a better world

The level of effort expended to gain situational awareness is driven by the needs of the recipient, from national command authorities down to the on-scene tactical commander of a small unit, be it on sea or land. The same applies to commercial entities, from corporate boardrooms to the hands-on manager at the lowest level. Ships today are larger, carry more cargo and have more advanced operating systems. Unfortunately, they ply the waters in growing numbers with increasing opportunities for exploitation and criminal acts. More than 90 percent of global trade moves on the oceans, transported in millions of containers on 50,000 merchant ships, flagged by approximately 150 nations. The ability to comprehensively monitor activities to identify trends, differentiate irregularities, and examine anomalies to support effective decision-making is a daunting challenge that benefits tremendously from multilateral cooperation.

A 2014 article devoted to how geospatial intelligence is assisting in maritime situational awareness in Trajectory, the official magazine of the United States Geospatial Intelligence Foundation, sums up the thrust of this paper very well.

> Although maritime domain has evolved consider-
> ably in the past decade, gaps remain. Filling them
> will require a sea change in three principal areas: data
> collection, processing, and *sharing*.

This chapter is both a discussion of where space-based global maritime awareness is going and a guide to what needs to be considered for those wanting to create systems that will help us get there. It is meant to be a self-contained document, so it will repeat some of the information presented earlier as it is the basis for this examination of the future of space-based global maritime awareness. It will focus on the sensor and data fusion elements to consider when building a concept of operations for space-based maritime situational awareness. I do this because I have now been in enough different situations where a great deal of time was spent (wasted?)

trying to get others to understand what advanced technology was available and needed to be included in our concepts of operations. Many people are still lost in the past.

Let us begin our look into the future by looking at the immediate past.

By the early 21st century it is widely recognized that there are four global commons where all nations of the world both meet and interact. Thus all countries have a stake in ensuring order in them. Those four commons, in chronological order of creation, are Maritime, Air, Space, and Cyber. In the last few years, more and more people are beginning to fully comprehend the fact, the leading-edge technologies of three of those global highways, Maritime, Space, and Cyber, are rapidly converging in the creation of a substantially improved Maritime Situational Awareness (MSA) with potential benefits for all of mankind by the creation of Global Maritime Awareness.

The immediate past was spelled out in previous chapters. This chapter will lay out why this is happening and how unclassified space systems are impacted by new capabilities from artificial intelligence (AI) coupled with machine learning and the rapidly expanding Internet of Things (IoT). Technology is substantially increasing maritime situational awareness for the good of any nation that employs it and, indeed, for the entire globe.

The methodologies used to achieve the required level of MSA involve substantial interaction in three of the four global commons. It should be made clear that space-based MSA will not replace the traditional terrestrial means of attaining MSA. Instead, it will make the conventional systems substantially more effective and efficient by capitalizing on the paradigm shift that has occurred in technology in the first decades of the 21st century, especially in regard to unclassified space systems, data fusion and analysis technology.

The last few years have seen explosive growth in the number and capabilities of Earth observation satellites, and most have at least tangential capabilities in the maritime domain. Space and data fusion systems using artificial intelligence and machine learning have become very cost-effective adjuncts and, when intelligently applied, can make traditional terrestrial systems dramatically more effective, thereby saving substantial amounts of money.

While it remains to be fully established how far these savings will outweigh the cost of employing the space systems, many of us who have studied the subject in depth believe this will be the case and that the savings and improved efficiencies will more than pay for the space systems by a significant margin. This hypothesis should be examined by the naval and national think tanks of the world as a matter of the first order. The payoffs should be huge.

The methodology to study the subject—and determine how to build upon what already exists—are basically the same. This final section could be used as a guide for both.

Accurately defining the problem you are trying to solve with your proposed system is key. Knowing where you want to go before you embark on a trip is particularly important if you are going to do more than just wander around for the fun of it. The United States' National Strategy for Maritime Security, September 2005, starts this way:

> The safety and economic security of the United States depends upon the secure use of the world's oceans. … Maritime security is best achieved by blending public and private maritime security activities on a global scale into an integrated effort that addresses all maritime threats.

These words apply equally to all maritime nations. Countries recognize that their economic well-being depends on access to the seas. But security on the seas, safety, and the protection of the maritime environment and its resources require vigilance on the part of each nation individually and, where possible, in conjunction with all countries of its region. This is because no one nation or business enterprise has sufficient surveillance resources to provide the required level of maritime situational awareness in all areas that affect its well-being, safety, safety, and environment and resource protection.

Many experiments across the global maritime domain have shown that nations working together on these issues are much more effective than single nations working alone. This is probably nowhere truer than in the application of space systems, a global capability by its very nature, to the maritime domain. Space systems are now being applied to provide dramatically increased maritime situational awareness in several locations across the globe, but there is much more that can be done, both individually and collectively.

Surveillance from space

The United States Navy, Coast Guard, and Customs and Border Protection use many different systems to achieve and maintain awareness both in our home waters and the approaches to our shores, as well as overseas. Watch centers at San Diego, Alameda, and Riverside, California; as well as Norfolk, Virginia; Suitland, Maryland; and Key West, Florida maintain watch in areas close to the United States. The Navy also has watch centers in Japan, Italy, Hawaii, and Bahrain. The United States also participates in the National Maritime Information Centre in Northwood, England.

These centers routinely use both classified and unclassified maritime surveillance sensor systems. Because these centers all operate at high classification levels, the managers of these sites are, understandably, reluctant to discuss specifics of which systems are used and how. Managers confirm they use both commercial terrestrial and satellite AIS. But they are less willing to describe their usage of space-based imaging systems at any level.

However, for the processing, fusion, analysis and display/ decision aide portion of operations they do use either their own developed software for what I labeled "static/dynamic data analysis." I cannot know what is happening in the classified world, but in the commercial, unclassified world the software tools in use are largely derived from pioneering systems such as Computer Assisted Threat Evaluation (CATE) that was developed by merchant Captain Jatin Bains and, more recent not long afterwards a system developed by Paul Kerstanski called Computer Assisted Maritime Threat Evaluation Systems (CAMTES). In Europe both e-GEOS and CLS developed similar systems, but they all bear a remarkable resemblance to CATE. It may well be a case of folks looking at the same problem and proceeding down a very similar path given the state of the art, but the resemblance may not be a coincidence.

The current US government for-official-use-only system is called SeaVision, and by early 2020 it was being shared with 84 different nations. It is clearly an amalgamation of a system developed by the Navy and US Department of Transportation based on Kerstanski's design, which was inspired by Bains'. Kerstanski subsequently worked for Bains, and their commercial tool is called Space Eyes. It is the direct descendant of CATE. By many reports it is a better tool than Sea Vision. Most of the static/dynamic data analysis tools are designed to take inputs from multiple sensor types as well as from numerous relevant word-based relational and informational databases. These systems are crucial parts of the overall effort and will be discussed in more detail later in this chapter.

Imaging systems, optical and synthetic aperture radars, also play a significant role, and at least one watch center director has said publicly that when the occasion requires it, they ask the National Geospatial-Intelligence Agency, part of the Defense Department, to provide coverage of a specific area. It is up to the agency to determine which system or systems to use. Initially, there were two programs: the EnhancedView and the Maxar Global Enhanced GEOINT Delivery—designed to allow it to quickly task commercial imaging systems providers such as Maxar/Digital Globe, Planet Labs, and Airbus plus others for optical coverage. Synthetic aperture radar satellite coverage would be tasked to DLR, e-GEOS, Airbus, MacDonald. Dettwiler & Associates, ICEYE and Capella Space. At least three other companies are now developing SARsats as well: Surrey Space Systems, PredaSAR and Alpha Insights. Unclassified ELINT satellites, the most recent Earth observation system, such as Unseen Labs, HawkEye360, Kleos and Horizon Technologies also now have similar tasking arrangements.

As one descends the strategic-to-tactical continuum, the demand for timeliness increases. However, one size does not fit all. In between are several strata of needs, from strategic down through operational to tactical. Thus, the concept of operations outlined here will not try to be all things to all men, but rather to lay out guidelines and things to be addressed. If adopted and practiced, these salient points should substantially increase maritime situational awareness—and thus maritime security— for all who practice the concept laid out in the following pages. It should also provide savings in the operational costs of both patrol and reaction forces by making their employment much more focused, and thus more efficient and effective.

Any truly useful concept of operations is a fluid document, capable of growing and changing with the times and present conditions. Accordingly, the concept outlined here will not be provided as struck from granite, steel, aluminum, and silicon, but rather a set of facts to help all guide development and deployment of systems for the good of both their nation and, potentially, the good of all nations involved with the maritime domain, and all peoples of goodwill. It also recognizes that each individual situation is best addressed by the country, or group of countries in the region, or the commercial organization or organizations affected, so it will not try to reorganize local entities.

Indeed, I encourage the development of regional organizations to develop synergistic arrangements in which the regional groups collaborate in the use of the satellite-based maritime situational awareness systems to be described in the following pages. It is probable that there is a need for a worldwide organization to coordinate the effort on a global scale. But there is a clear need for regional cooperation, whether or not a worldwide organization evolves.

These space systems, under-used in the maritime world until very recently, are a new, and until recently, mostly unappreciated, set of tools that could be used to aid in substantially improving maritime situational awareness at all levels. The one place where their capabilities are being widely used in Europe, probably the one place on Earth with the smallest need. But even there, the requirements are also huge.

Warfare in the fourth dimension

Numerous models strive to depict the process by which an entity is detected, categorized, identified for a response, and then dealt with appropriately. This ranges from being ignored as being completely legal or not being in any danger or not being worth the effort to prosecute at one end of the scale, to needing to be destroyed as soon as possible as a deadly threat involving weapons of mass effect, and everything in between. The model that will be used here to examine and explain the space-based maritime situational awareness concept was developed almost 40 years ago, at the US Naval War College to support the work of the very first Strategic Studies Group and stresses the requirement for sufficient time in every level of encounter, from strategic to tactical. At all levels of command, time to understand the situation, including one's options dictated by force availability, national guidance such as rules of engagement, the tactical situation, etc., is the critical component of success. Hence the name of the model: Warfare in the 4th Dimension.

The elements of the model consisted of:

1. Task (to determine which sensor or sensors are best suited to collect the desired data);

2. Sense (to collect the data). In the cyber world, this could be an automated data search/data, semantic fact gatherer;

3. Process (normalization of the data collected from both sensors and word-based (semantic/static) data sources to allow insertion into a software program to permit fusion with other data and information);

4. Fuse (to marry the data with similar and dissimilar data and information to understand its unique attributes);

5. Analysis (of the results of the fusion);

6. Display and Decision Aid interaction (with the results of the Fusion and Analysis). This is a critical step because a well-designed display system is essential for both the analyst and

the command authorities to understand the tactical situation quickly. It is also vital to have both the analysts and command authorities working off common data, commonly displayed, (this is where static/dynamic data analysis tools such as SeaVision and Space-Eyes fit in);

7. Disseminate (both the final product and orders to all affecting parties, from strike/interdiction units, which are to engage the target with the appropriate level of force, to orders to unarmed or overmatched elements which will need to move out of harm's way);

8. Action (as directed).

We are going to discuss only the first six elements of this model, leaving Dissemination and Action to organizations to plan on using the means and forces at their disposal. However, builders of concepts of operation must keep in mind the capabilities of the forces at their disposal. To do otherwise is a quick way to guarantee failure.

Many nations now recognize these facts and are taking steps to increase capabilities in Maritime Domain Awareness and Maritime Situational Awareness. Countries are collaborating on developing requirements for technology for ships, data systems, and information-sharing. Prioritizing information collection is not new, though the focus on maritime awareness has expanded considerably since 9/11, and become an increasingly higher priority. This is probably due to two facts, the growing awareness of the importance of the maritime domain to the well-being of the world, and significant advancements in marine surveillance systems, especially both terrestrial and space-based detection of AIS and other emitters via the new unclassified ELINT satellites constellations.

Breaking it down

The first element of this concept is the tasking of the most advantageous space system. Things to be considered include the following: capabilities of the spacecraft; availability within the needed time frame; time to upload taskings, including location of uplink sites; time to move to collection opportunity/window; time to downlink opportunity/time to downlink; time to process the data collected; and time to disseminate.

There are at least two tools on the market today that meet all these needs: SaVoir of Taitus Software; and Orbit Logic's Collection, Planning and Analysis Workstation. Space Eyes also has this capability.

The second element of this concept is sensors: The most crucial sensor in many instances is the one that does location and tracking. The primary means of tracking maritime traffic in coastal waters is, of course, radar. But terrestrial radar is limited to line of sight, roughly 25 miles versus a ship, so as space technology has evolved it has been employed in tracking vessels.

Argos was the initial ship-tracking system specified by many countries as part of their Vessel Monitoring Service of their fishery management efforts. The monitoring, control and surveillance element of the vessel monitoring system was developed in part to counter illegal, unreported, unregulated fishing. It was one of the first uses of space systems beyond research, communications, and weather reporting.

The Argos system, a collaboration by the French Space Agency, the National Aeronautics and Space Administration, and the National Oceanic and Atmospheric Administration, was originally a scientific tool for collecting and relaying meteorological and oceanographic data around the world.

In 1986, the French Space Agency, CNES, created a subsidiary, CLS, to operate, maintain and commercialize the Argos system. Currently, several other international space agencies also actively participate in the Argos system, including the European Organization of the Exploitation of Meteorological Satellites, the Indian Space Research Organization and others. Since then, the Argos system has been used to systematically track ships, primarily fishing vessels, as part of the vessel monitoring system (VMS). Since that time, a number of other satellite-based ship location systems have been developed, for a variety of reasons, but VMS remains a primary need due to illegal fishing.

Any vessel tracking system, including VMS, requires technology on the vessel, ashore, and communications between them. Also, there may be additional communications from the Fisheries Management Center of the vessel's country of registry, and regional or national centers of the waters in which the vessel is fishing.

The most basic function of a VMS is to determine the vessel's location at a given time, and periodically transmit this information, to a monitoring station ashore. Different systems use different communication technologies, including AIS via ORBCOMM, Inmarsat, Iridium, and Argos, depending on the functionality required by the particular VMS system.

VMS components on the vessel sometimes are called VMS, or sometimes Automatic Location Communicators. These minimally include a GPS antenna and receiver, a computer (which may be embedded or user-supplied), and a transmitter and antenna appropriate for the communications that links the vessel to the flag center. In practice, many of the VMS components also have applicability, along with non-VMS marine electronics, to a wide range of functions aboard a fishing vessel. These include navigation, fish finding, collision avoidance, routine voice, and email communications, etc.

Every one of these systems, including AIS and S-AIS, require the installation of additional equipment, but S-AIS has a significant advantage in that AIS's primary function is that of a collision-avoidance beacon, now a required installation in all ships over 300 tons engaged in international trade, all passenger ships, and all tugboats over 600 shaft horsepower. Additionally, due to its relatively low cost and high utility as a collision-avoidance beacon it was initially designed to be, the use of AIS has migrated to many other vessels not required by law or regulation to carry it. Most fishing boats of significant size, yachts, and many pleasure boats, as well as working boats now carry AIS as a safety device. That being the case, we will spend some time describing AIS and S-AIS in detail.

The international confederated collection on a near-global scale of the Automatic Identification System (AIS) such as the Maritime Safety and Security Information System has proved to be incredibly useful for maritime safety and security purposes. MSSIS is a government-to-government AIS data exchange developed by the US Navy from a Department of Transportation system developed for the US's Maritime Domain Awareness Program Integration Office for use as a security measure for the Republican and Democratic national conventions of 2004. MSSIS was first deployed in the Mediterranean in 2006 and has proved so helpful it now involves countries from all corners of the world.

As was discussed in the chapter on satellite AIS, ships broadcast the AIS signal every few seconds, with the interval decreasing as the speed increases. The ship's latest positional data is updated automatically and routed into the AIS transmitter via an interface from the ship's systems. But when a vessel is transiting the high seas or in a remote area, until the advent of S-AIS, this data could not be picked up and used by shore-based security centers. That has changed. Many maritime watch centers, both governmental and private, now routinely couple AIS data with inputs from a wide range of data sources. These sources are as varied as ships' status reports, brokerage records, published ship movement records and current reports, historical data derived from now over 20 years of AIS collection, with data analysis tools such as Computer Assisted Threat Evaluation, or its newer competitor, Computer Assisted Threat Evaluation System, and now SeaVision and Space-Eyes.

A ship with an operating AIS beacon automatically provides significant information to all similarly equipped vessels or shore stations within range such as the vessel's name, location, heading, and speed. AIS operates in the VHF maritime band, so the range is restricted to line of sight, approximately 25 nautical miles, depending on antenna height. AIS has 29 data fields, although not all of them are used by all ships. Information broadcast every few seconds includes ship name, location, course, and speed. The complete set of all 29 data fields, which includes such things as flag state, length, beam and draft, destination, and the vessel's last port, is broadcast every five minutes or on request from a base station. The information is unclassified and thus easily shareable among organizations with differing levels of security access, including multinational groups. A key benefit of AIS is its ability to operate in all conditions, including periods of reduced visibility.

AIS does not supplant radar and camera surveillance systems, but it does make them much more useful. The same is true of the now available space-based surveillance systems that operate globally and are especially valuable for the open seas, beyond line of sight from shore, and other remote areas where terrestrial radars and cameras do not reach because they are limited to line of sight. However, since the 2008 introduction of the Satellite-AIS constellation by ORBCOMM, all that has changed. S-AIS now provides near-universal collection and reporting coverage.

The middle four parts of the operational concept: Parts 3 (Processing), 4 (Fusion), 5 (Analysis) and 6 (Decision/Displays Aides) are clearly different functions but, as outlined earlier in this chapter, these separate functions are often combined in one software tool. There are a number of tools in use, but the main ones are Sea Vision for governments and Space Eyes, developed by Channel Logistics. They developed the first such tool, starting in 2003/2004, and continue to be a leader. Many of the sensor companies also have their own tool that does much, if not all these functions.

Chapter 12

A Flourishing Technology

Tactical as well as strategic

As previously discussed in detail, AIS, which came into use in the early 2000s, has been made dramatically more useful with the launch of satellite-based AIS receivers, in a six-satellite constellation, by ORBCOMM, for the US Coast Guard in 2008. S-AIS was designed to have two primary uses.

The guided-missile destroyer USS Bainbridge *tows the lifeboat from the Maersk Alabama to the amphibious assault ship USS* Boxer, *in background, April 13, 2009, to be processed for evidence after the rescue of Capt. Richard Phillips. Indian Ocean pirates off the Somali coast had held Phillips hostage in the lifeboat. Photo by Marine Lance Cpl. Megan E. Sindelar*

The first goal is providing tactical and relevant data, in a useful timeframe, by identifying and tracking all ships in any location on Earth. This information can be automatically parsed to allow a law enforcement or intelligence agency to focus attention on geographic areas. Thus, watch centers can know where all compliant legal ships are in their area of regard. That information can help a government determine where a synthetic aperture radar satellite should be tasked to image specific areas of interest at sea. SAR sats can detect even small targets down to less than a meter as well as wakes of vessels. Indeed, the automated processing of these images has been refined to the point where, instead of sending the imagery—which requires a trained analyst—reports are all that are sent when pre-set reporting conditions are met. For instance:

> Unknown probable commercial bulk carrier detected at 42-12.4N 172-15E. Approximate length 190m. Approximate heading 205. Approximate speed 18.5 kts.

The second primary use of S-AIS is more strategic and operational but can directly influence tactical operations. It has been known for many years that tracking vessels over an extended period in one's area of responsibility allows for the development of pattern analysis (the patterns of life). Thus, once you know what is normal and can readily recognize anomalies. AIS, both terrestrial and especially satellite, provides for the easy collection of all legitimate traffic over an extended period and the development of in-depth patterns of normal operations. The study and analysis of S-AIS track and identification data over time provides a baseline of what ships are normally operating in specific locations.

AIS has this vulnerability: All a ship has to do to not be tracked is to turn its AIS off. However, most ship masters operating legally are very reluctant to turn off their beacons due to legal liabilities in the event of a collision while the beacon was off. Indeed, on commercial vessels over about 100 feet in length, AIS data reports are usually automatically routed directly to a radar scope which displays the received data. The system is equipped with software that automatically alerts when another vessel is on a collision course or is even just getting dangerously close. Many smaller vessels have this function, too. Masters engaged in legal activities welcome AIS as a never-sleeping lookout over other legal large ships even at night, in fog, and in the worst weather conditions.

Algorithms have been developed by several organizations, most notably the European Commission's Joint Research Center, the Italian National Institute for Environmental Protection and Research, the Israeli software company Windward, and the US company Space-Eyes. They define "normal" traffic and conditions, such as regular shipping traffic, as well as "abnormal" situations. That might include a rendezvous of vessels or small, fast boats headed to an unusual location. The report can help the tactical commanders decide whether to respond by sending a vessel or aircraft. And it may enable a businessman to make an informed decision.

In late 2003, the US interagency Maritime Domain Awareness Program Integration Office, with the assistance of Johns Hopkins University's Applied Physics Lab, commenced investigating the feasibility of satellite AIS. In early 2004 the office made the decision to use available Coast Guard money to fund the development of satellite-based AIS. The effort, in which I was deeply involved, culminated with the 2008 launch by ORBCOMM, the satellite network company based in New Jersey, of the first operational constellation of six satellites. These satellites provided a global collection capability for AIS transmitting ships several times a day, significantly improving awareness.

As more S-AIS systems were placed in orbit and constellations of specific companies have increased in size, there has been an accompanying, and dramatic, decrease of the time interval between collection and availability for users. This interval will continue to decrease as more S-AIS receivers are placed in orbit (28 were operating as of December 15, 2014 and as of early 2022 there are about 205) more ground stations are deployed globally to receive their data. ORBCOMM, placed its 17th Earth ground station in use in 2016, and completed its 19 satellite Next Generation constellation in December 2015. Coupled with its existing satellites, the advances meant that the time from collection to report was less than five minutes for most parts of the world. Indeed, in the mid-latitudes, the time was down to less than two minutes.

ORBCOMM, one of the world's largest suppliers of machine-to-machine (M2M) communications, the backbone of the Internet of Things (IoT), has a robust constellation now operating—just one of several S-AIS systems that presently operate. Spire, the global data and analytics company based in San Francisco, announced plans in 2014 to build a constellation of 25 "two-liter sized" AIS satellites.

As of early 2022, Spire has far surpassed its initial plans launching over 100 AIS equipped spacecraft. Where and how it downlinks the information from these satellites has not been discussed in public but is critical in determining the tactical usefulness of its constellation.

The Canadian company exactEarth, the other main player in this business, partnered with Harris Corporation—which would merge into L3Harris Technologies, and had nearly 60 AIS receivers on the Iridium NEXT satellites. Importantly, these satellites use the Iridium crosslink to downlink the AIS signal in near real time to its processing sites, thereby allowing very rapid (< 1 minutes) delivery of AIS location data to its customers. In early 2022 Spire acquired exactEarth and they now have the largest S-AIS constellation by a significant margin. Going forward, it is more likely that the question will be how often a report on an individual vessel is needed, rather than whether such information can be obtained.

Space-based Global Maritime Awareness came into being with the ORBCOMM 2008 launch. But Global Maritime Awareness was not really complete as a system until the launch of unclassified radio frequency (aka electronic intelligence or ELINT) satellites in late 2018. HawkEye 360's Pathfinder is the first RF geolocation satellite to be made public. Unseen Labs of France actually beat them into orbit, but have remained much more close-mouthed about their system. Kleos of the UK and Horizon Technologies, also now have entered this market and all seemed to have lived up to expectations. I suspect these systems will be the first of many.

However, the term describing the new satellites is a bit cumbersome. Indeed, I originally called them unclassified

ELINT (electronic intelligence) satellites, but some folks were uncomfortable with that label, even though Soviet's ELINT satellites were discussed and described in many open sources starting in the mid-1980s, and that is what they are. Just to be politically correct I suggest if we cannot call them ELINT sats, we call them RFgeoSats, but I am open to suggestions. You saw it here, first.

It has been recognized for some years that RF geolocation (ELINT) would be an extremely useful tool for maritime awareness, especially when used in collaboration with S-AIS. It fills a need to track ships when they turn off their AIS, as many bad actors now do when they commence nefarious actions such as smuggling or illegal fishing or, perhaps, pumping their bilges in restricted waters. But you still need AIS to identify the ships on initial contact, before they turn it off. These two systems, S-AIS and ELINT Sats are not in competition with each other as some suggest. Rather, they are synergistic. We originally did not have enough firm data to be absolutely sure, but now it is clear.

Another tracking system now mandated for global use is the Long Range Identification and Tracking (LRIT) reporting requirement of the International Maritime Organization. LRIT requires ships to broadcast their position every six hours via their satellite communication system already required by IMO's Safety of Life at Sea Convention. Generally, this requirement is met by ships setting their communication system to report its position to the correct LRIT address automatically at the proper time interval.

The main difference between satellite AIS and LRIT, a question that is often asked, is that LRIT requires a conscious act by the master of a vessel whereas S-AIS collects its data from all ships broadcasting the signal as the collision avoidance system it was designed to be. Also, AIS is designed to exchange position reports with every other vessel within line of sight that is also equipped with the AIS as a collision avoidance beacon, while LRIT reports only four times a day, and then only to a national reporting center, not to other ships in the area. With both AIS and LRIT, data may be manipulated and misused.

ARGOS/VMS, AIS, S-AIS, and LRIT all provide the important location information complementing other data sources and multilateral collaborations to secure information, including air surveillance, port security entry requirements, human intelligence, radar, video cameras, and patrol craft. Terrestrial AIS and S-AIS, as well as LRIT, complement national-level organizations and private organizations such as IHS Fairplay to fuse, analyze, and disseminate information.

When bad actors don't flip the switch

S-AIS is especially useful when paired with the several radar satellite constellations now operating. S-AIS also allows optical imaging satellites to be significantly more effective over the oceans. S-AIS offers both systems, radar and optical, a realistic chance of understanding who they are imaging and, if the imaging systems detect a ship that should be transmitting and it is not, then questions are raised, and resources can be focused on that dark ship. The collection of AIS from both terrestrial and, especially, satellite systems, also provides the opportunity to develop MSA shipping pattern analysis, a handy tool for assisting in the detection of operational anomalies.

Several nations operate satellites with synthetic aperture radars. The German Aerospace Center, MacDonald Dettwiler in Canada and Rome-based e-GEOS operate commercial ventures in conjunction with their national missions selling imagery. ICEYE, a Finnish microsatellite manufacturer, has an excellent small satellite with amazing capabilities, given its size. Capella Space, in San Francisco, recently joined this group. The clarity of its early images was impressive. Space Alpha has announced they will be placing a dual-band SAR satellite in orbit next year with extremely fast processing that will allow onboard processing of the SAR data, a technology first.

After the imagery is collected, it is placed in an archive and, with some restrictions, is available for sale. For a fee, these organizations will take an image of a specific area at a specified resolution in a requested time frame. Other entities are considering following suit.

The same is true of optical imaging satellites systems, but let us stay focused on the SARsats a bit longer as they are now a critical part of marine situational awareness. Their importance is sure to grow as their numbers and capabilities increase, as they are sure to do, and the cost per image goes down, as it is almost sure to do. All SAR satellites can take images day and night in most weather conditions with at least three levels of resolution, from wide swath/low resolution to medium swath/medium resolution to spotlight coverage of a small area with very high resolution, now approaching resolution levels (< 1 meter). That resolution was not possible even for optical systems 20 years ago. The advantage of SARsats over optical imaging systems is their ability to take their images day or night, rain or shine, while optical systems need daylight and clear skies.

While the SARsats have been improving their capabilities, the optical images have been working hard to improve their systems as well, and they can now take images of significantly better than one square meter and they certainly do have a place in marine situational awareness. The three major optical imaging systems are operated by San Francisco-based Planet Labs with over 220 satellites in orbit, Colorado-based Maxar Technologies and systems of the European partnership of Airbus and Thales Alenia Space. These last two companies operate over a dozen optical imagers between them, and their products are readily available for sale. Indeed, the National Geospatial-Intelligence Agency has a contract in place that allows both quick tasking of the Maxar systems if the need for imagery is urgent and a means to process and disseminate the finished product quickly. This is especially useful if the requirement is for an unclassified image that can be shared in such events as humanitarian relief or disaster preparedness.

Airbus Defence and Space's one-of-a-kind constellation of optical and radar satellites provides access to any point on Earth at least twice a day. They have more than 30 years of production expertise as well as direct access to the satellite sensors via their global ground station network, facilitating rapid data acquisition over crisis areas and fresh imagery for regular updates. The entire constellation can be tasked via a desktop application.

As good as the optical systems are, it is the simultaneous collection of S-AIS and SARsat images that has the attention of people worldwide. Spain saw the launch of PAZ, its first spy satellite, in 2018 and has both an excellent German-designed, x-band high-definition radar and at least one AIS receiver. Canada's next generation AIS, the Radar Constellation Mission, is already approved and funds have been identified; it will consist of at least three SARsats, each equipped with AIS receivers.

NovSAR, built by Surrey Satellite of England and launched by India in 2018, also has both a SAR and AIS onboard. Italy is considering fitting the next generation of its excellent Cosmo SkyMed SARsats, which currently consists of four SARsats in a single constellation with basically the same orbital characteristics, but with the timing staggered by approximately 15 minutes (+/-). Thus, they can cover wide swaths of the ocean quickly. The main problem with this system, as with all SARsats, is that they are power limited. They can only transmit about 10 percent of the time on each orbit of Earth. SpaceAlpha, a Canadian company is in the process of building a dual band (L and X) SAR with AIS onboard. It is planned to be able to collect AIS and wide area search with its L-band SAR and process the fused SAR and AIS information onboard. If a target of interest is developed, the satellite will automatically immediately, on the same pass, image the contact with its high resolution X-band SAR in spotlight mode for further analysis.

Besides SAR and optical satellites, high frequency surface and skywave systems have some capabilities beyond line of sight/over the horizon collection. But they are costly to install, man, operate and maintain, and high frequency surface wave transmissions can interfere with critical terrestrial emitters.

The European Maritime Safety Agency and FRONTEX, the European Union's Board Patrol and Coast Guard, are the foremost users in the world of commercial space systems for marine situational awareness. in the world. They initially commenced their MSA work using periodic SARsat images looking for polluters where ships were known to illegally pump their bilge water to sea. They used LRIT to determine the probable identity of vessels that appeared to have done something illegal. After a very few years, the European agency commissioned the French company CLS Group to conduct a cost/benefit analysis of replacing LRIT, which they got for free, with S-AIS, for which they would have to pay.

The results clearly indicated that the more costly S-AIS was well worth the price, and the European Maritime Safety Agency has started using it as their primary ship-tracking tool. The agency is one of the strongest proponents for S-AIS. The SARsat detects the offending act; S-AIS tells them who it is. Both systems, SARsat and S-AIS, are crucial to mission success.

For some types of illegal activity, such as smuggling contraband, drugs and perhaps people, use of new technology might not be quick and straightforward. It may take days or even weeks of studying for patterns revealed by weeks or months of S-AIS collection. But patterns detected can give a watch center a much better understanding of where to point imaging sensors. Anomalies also can inform maritime authorities where to place terrestrial search and interdiction assets, and in what geographic areas to patrol.

Another way to maximize one's investment in MSA systems is the regional sharing of data with countries with similar interests and concerns. However, multiple types of data come in many different formats thus processing is required to normalize each kind of data to allow its fusion.

Data fusion is the process of combining data from multiple sources, including both sensors and semantic (word-based)/ static databases to create the most complete and accurate awareness and understanding of objects in the environment of regard as possible. Combining multiple sensors with greatly varying range and fidelity to detect different signatures a vessel emits (electronic, acoustic, visual) with word-based databases with such informative data as a ship's history as to ownership, operating organizations, past voyage data, crew and passenger manifests, can produce a precise and useful picture. Identifying the same track as seen by different sources is, in reality, a big challenge. But some algorithms can correlate multiple data sources, predict where a record is going, and pick out aberrational behaviors from normal traffic patterns. For example, combining the AIS data with SAR satellite ship detection to find anomalies is possible, but computationally intensive.

The arrival of Big Data

Very closely related to data fusion and another significant stride forward in MSA is the advent of so-called Big Data with analysis capability coupled with artificial intelligence and machine learning. With the advent of cloud technology, Big Data can be accessed from anywhere. An organization does not need a rack of servers and their computers. More than a dozen companies and organizations have exploited this new technology and are concentrated on improving on how data can be processed, fused, and analyzed to enhance MSA (with S-AIS as a leading component), among them Google, Vulcan, Analyze, Channel Logistics (DBA Space-Eyes) and Spire in the United States, and CLS (France), DLR (Germany), MDA (Canada), and KSAT (Norway).

The US Department of Homeland Security's Science and Technology Directorate has created a fusion center tool, originally called the Coastal Surveillance System. They use a system-oriented-architecture where all available data can be gathered, analyzed by one of the dynamic data analysis tools mentioned earlier, then disseminated to all who have the correct permissions. Very importantly, the information will not have a Defense Department or Intelligence Community classification so it can be distributed much more widely than heretofore.

Although I had a fundamental disagreement with the initial title of this program tool, the Coastal Surveillance System was among the most exciting technologies on the horizon. The objection that this author has is that this tool was called a *coastal* system when it has the clear potential capability to be a *global* system. That is a good problem to have. The reverse, to call a system a system *global* when it clearly had only a *coastal* or *regional* capability, would be much worse. The coastal system has been renamed the Integrated Maritime Domain Enterprise

System with plans to transition again into the Integrated Domain Enterprise/Awareness System.

The excellent functionality concept remains the same, just the name changes. I am most interested to see this tool in use, given that I been following its development for something like 14 years now. No one ever said this was an easy task. The key point is that it is planned to incorporate all sensors, data, intel sources, including Space based data. Sounds almost too good to be true, but we shall see.

Summary

Maritime Domain Awareness or Maritime Situational Awareness, or Global Maritime Awareness, no matter what we call it, is more than collecting and analyzing data. It is necessary to determine if the information analyzed reveals a need for action, and how best to act on it; then acting on it. I describe the process as the Task-Sense-Process-Fuse-Analyze-Decide/Order-Act cycle. We have discussed the sense-to-decide part of the cycle but avoided the last two parts of the cycle as they are reserved to the operational commander and are beyond the scope of this chapter.

Space systems already play a significant role in the entire sense-to-act cycle that is the heart of MSA. Their role could be much more significant; indeed, it could be global. But let us take first things first and develop the existing capabilities I've described into an effective national system, realizing that, in order to have a genuinely effective national system, it would be of substantial assistance—if not an imperative—to have international information sharing, and collaboration.

Pertinent technologies are rapidly expanding on many fronts. This includes: new and upgraded sensors, with significantly increased ability to collect; high-capacity processors that can rapidly sort and provide understanding of the received data; and the ability to both quickly transmit vast amounts of data and to have multiple locations simultaneously working on the same set of data. Space systems already play a role, but this author believes that role is bound to expand exponentially in the next few years.

In the Crawl-Stand-Walk-Run model we are at the *Stand* level. But people all over the world are beginning to take the first steps in this new direction. They will be breaking into a *Run* soon.

I stand ready to assist in the creation of such a system for any legal organization, any nation, or group of nations. The need is there, the capabilities are there. We just need to put them together in a smart way.

Appendices

Herewith are my selected writings on Maritime Domain Awareness/Global Maritime Awareness over a two-decade period starting in late 2001. They are meant to provide insight to researchers as to how the concept evolved from a maritime traffic tracking system through Satellite AIS to space-based Maritime Situational Awareness to "Collaboration in Space for International Global Maritime Awareness (C-SIGMA)" to "Global Maritime Awareness." I hope this will be useful in understanding how this game-changer in maritime security took shape.

An example of the European Space Agency's radar satellite imagery combined with S-AIS data ship tracking on May 8, 2019. Courtesy of author

Appendix One
A Maritime Traffic Tracking System:
Cornerstone of Maritime Homeland Defense

Author's note: What follows is a slightly revised version of an article that appeared in the Naval War College Review, *an article that evolved from a paper I wrote after 9/11 in response to President Bush's task to the Chief of Naval Operations. That paper was sent to the many participants in the two wargames and the symposium/workshops we held in the first few months after 9/11. I sent drafts of my work to literally hundreds of people, many of whom responded with comments. This paper was my distillation of that earlier, classified work. My aim was to provide for the continuation of the focused, informed debate on what, why, and when we should build a maritime equivalent of NORAD. Some of the comments I received have been incorporated and many others led to more research, thus refining the paper, and transforming it to a better product. Dishearteningly, some of the comments clearly indicated the reviewers had not read the paper, but had their own ax to grind, pet project to push, or prejudice to vent which I identified earlier in this book. Those I generally ignored.*

Among the many lessons 9/11 has taught us is that we are a vulnerable nation. This is especially true on our sea frontiers. President Franklin D. Roosevelt understood this and made a point of it during his first fireside chat after Germany invaded Poland, plunging Europe into war in September 1939, 27 months before the United States Navy was attacked at Pearl Harbor. American security was (and is) "bound up with the security of the Western Hemisphere and the seas adjacent thereto. We seek to keep war from our firesides by keeping war from coming to the Americas." Today, we are engaged in a different war, and war has already come "to our firesides." To help prevent its return we must again attend to the security of the seas and our ports. This is doubly true for, despite the emergence of the information age and the decline of our merchant marine, we are still a maritime nation and the security of our harbors and seaports is still of first importance to the well-being of this country. We are very dependent on maritime trade as was recently demonstrated by the significant economic damage done by the truly short dock strike on the west coast. It is easy to envision the economic cost and social impact of simultaneous terrorist attacks on two or more of our ports.

To maintain the security of our ports, the United States needs to track and identify every ship, along with its cargo, crew, and passengers, well before any of those ships, and what they carry, enter any of the country's ports or pass near anything of value to the United States. This article proposes a system that would provide that tracking capability, as well to meet any emergency with an appropriate response. A result of several months of wargames, conferences, and working groups dealing with the maritime aspects of homeland security, this article is intended to be a thought-starting strawman, a means to generate informed debate on how and why we may want to build a maritime counterpart to the North American Aerospace Defense Command (NORAD) and the Federal Aviation Administration (FAA) flight-following system. Drafts of this paper have been circulated since November 2001 to stimulate informed debate and information exchange. That information exchange has resulted in this paper going through several major revisions. However, more public discussion is needed until the concept and procedures outlined here are fully implemented. One disclaimer is appropriate. Though this paper identifies some specific systems to provide points of departure for further investigation, it is not intended to champion any specific system or systems.

Some people do not support the idea; they believe it is too hard and/or not worthwhile. However, Admiral Vern Clark, the Chief of Naval Operations has twice called for the creation of a "maritime NORAD." He first urged its creation on 26 March 2002 during a conference at Cambridge, MA on homeland security issues, sponsored by the Coast Guard and the Institute for Foreign Policy

Analysis. Parts of his speech resemble an earlier version of the paper that is also the genesis of this article, which had been forwarded to his staff. Though other powerful members of our government were among the speakers, it was Admiral Clark's words that the press highlighted. Admiral Clark's second call, on 15 August 2002, at the 2002 Naval-Industry R&D Partnership Conference in Washington, DC, also sounded to some as a synthesis of an earlier version of the white paper from which this article grew.

In conducting homeland defense, forward deployed naval forces will network with other assets of the Navy and the Coast Guard, as well as the intelligence agencies to identify, track and intercept threats long before they threaten this nation.

I said it before and I'll say it again today: I'm convinced we need a NORAD for maritime forces. The effect of these operations will extend the security of the United States far seaward, taking advantage of the time and space purchased by forward deployed assets to protect the US from impending threats.

This time the press missed the message. Some question what the admiral means by "forward deployed assets." If he is talking about units deployed overseas, the problem is significantly more difficult than if he means units deployed several hundred miles at sea off our coasts. This article describes how a maritime NORAD-like system could be built from existing technology to solve the twelve-to-1000-mile coastal belt protection problem. The overseas, far forward deployed problem is a closely related, but separate issue. Its multi-national political dimension alone makes it substantially more difficult. It is also significantly more difficult technically. The proposed system would assist in the far forward deployed problem, but we need to solve the problem of security at home first before we try to take it overseas. To try to solve the overseas problem before we address our own coastal problem would result in a huge waste of time and national resources, including both manpower and money. It may be possible to expand overseas the tracking capabilities required once they are in place in our coastal and economic exclusion zones, but it would be near impossible to establish the required tracking capabilities in foreign seas and then extend them back to the coast of the United States.

The problem

Every day more than 200 commercial vessels and 30,000 containers arrive at 185 US deep-water ports. The container carrying ships are concentrated in less than a dozen ports with the proper handling equipment, but most ports can accept a few containers. Additionally, every day hundreds, if not thousands of pleasure boats, fishermen, tugs with or without tows, offshore oilfield support vessels and research ships are active in the vast area from 50 to 1000 nautical miles off our shores. All these vessels are large enough to carry significant cargo. They sail not only to and from the 185 ports mentioned above but also to and from an even larger number of smaller moorings and anchorages. Some of these vessels, of all sizes and types, are involved in illegal activities such as drug and immigrant smuggling and/or environmental pollution.

The concern after 9/11 is that there may be other vessels with even more sinister objectives. This concern is heightened by the fact that tens of commercial vessels disappear every year. Some sink because of weather or unseaworthiness. Others probably "disappear" for insurance purposes; a few are pirated. Additionally, good-sized older, but still serviceable ships can be purchased in many places for less than the terrorists probably spent to execute the attacks on 9/11. Any of these vessels could carry enough explosives to destroy or damage substantially a port's infrastructure, including its bridges, chemical and petroleum plants, and processing, handling and storage facilities, and such high-value/high-payoff vessels as aircraft carriers and liquid natural gas (LNG) carriers. Indeed, the easiest way to put a weapon of mass destruction into large urban areas such as New York, Los Angeles or Hampton Roads is to send it by ship. This is especially true now that the airport security has been significantly enhanced worldwide.

Proposed solution and boundaries

These facts make it apparent that we need a better means than we now have of identifying and tracking all vessels, their cargo, and people as they approach the coast of the United States and its territories such as Guam and Puerto Rico. We do not have it now, but we need to create a system that will allow us to gain full situational awareness of all approaching vessels—their identification, position, course, speed, intended port of call, the identities of everyone onboard, and a description of the cargo and type of work each vessel is engaged in, just as is required for all aircraft, private and commercial alike. In other words, what we need is a required float plan the maritime equivalent of aviation's flight plan, and a means of identifying each vessel well before it nears our coast, the maritime equivalent of an IFF system. Moreover, we must be able to link the two, the float plan and the maritime IFF system, together. One proposed name for this new infrastructure is the North American Maritime Defense Command (NORMAD).

Let us see how we can do this. We can start by using several existing technologies to gather, process, analyze, and fuse data from all useful sources so those who must daily make decisions can reliably make the right decision and take appropriate action. As we have already noted, the proposed system centers on a maritime analog of the FAA's flight-following system. This is the critical initial step in building a maritime equivalent of NORAD. Though it does not adequately address the very real and difficult problems of tracking the cargo and people in or on the ships, this step will provide the needed information backbone with which the efforts others are making in those two areas can be combined. Indeed, the need to track the contents of a vessel—its cargo, crew, and passengers—is also a high priority and absolutely must be melded into this system. Though this paper focuses primarily on how we can create the required information backbone system, it also addresses the other issues to provide a context and to describe an end-to-end system of systems.

In conjunction with establishing the technical systems described here we must also establish and make widely known a policy denying any vessel freedom to approach our shores any closer than twelve nautical miles unless they have switched on and operating any one of the several maritime IFF systems described below. Moreover, the system they have must have been operating for at least 96 hours before entry into the coastal waters of the US. If the voyage from their port of departure takes less than 96 hours then they will have to have the system operating as they get underway, even before leaving port. Also, all vessels bound for U. S. waters should be required to file a float plan with the information detailed above, and get a registration receipt from the U. S. Coast Guard before they reach the 96-hour boundary. Those who do not comply should expect to be stopped, searched, and denied entry to our ports for a minimum of two days. The float plan could be forwarded via email, or fax, or any other record communications system. Most commercial companies already do something similar to this, just to keep track of their assets and maintain business flow. It is also a good idea from a personal safety view as a float plan tells someone ashore where a vessel is headed and when it expects to get there. If the vessel does not arrive in a timely manner, a search can be initiated. There have been two recent cases where single men sailing alone have spent more than three months each adrift at sea after their boats were disabled, primarily because no one thought to look for them. Many smaller ships operating offshore have some form of communications now that will support e-mail or fax. Those that do not can use a marina's e-mail or fax before departure. Whatever may be the vessel's or its operating location's facilities, it will be the operators' responsibility to make the necessary reports and possess the necessary documents. Given the widespread availability of communications systems, this requirement should not be arduous. The cost of the transponders for US citizens could be funded with an income tax offset. e.g., if one buys and installs one of the transponders/ transceivers the cost would be deductible from their income tax. This would also stimulate the electronics business, which would be an exceptionally good thing at the moment.

Vessels on voyages originating and terminating within US waters would need the proposed system only if they were venturing more than 50 miles offshore. Surveillance would be focused on the belt from 50 to 1000 miles offshore. The requirements spelled out in this proposal are consistent with international practice regarding freedom of navigation on the high seas. Indeed the US Coast Guard already has a 96-hour port entry notification requirement that dictates vessels broadcast their intentions well before they cross the 1,000 mile line. Once within 1,000 miles the proposed maritime IFF system would require vessels to update their position at specified intervals as they close the coast. The 50 NM boundary will eliminate from surveillance the vast majority of pleasure and fishing boats and other coastal commercial vessels that normally do not routinely venture far offshore. There are exceptions such as between Florida and the Bahamas, and off San Diego, where steady streams of boats are going both ways. Those few places will need special attention that can be provided with the establishment of radar identification zones.

Those areas are the six nautical areas that abut our neighboring countries' borders. They are where the coasts of Texas and California meet Mexico, where Washington state and Maine meet the Canadian coast, and the Strait of Florida, which abuts the territorial waters of Cuba, as well as the area of Puerto Rico and the US Virgin Islands. Radar surveillance in these six distinct areas would greatly facilitate the positive identification of all maritime traffic in these high interest areas. This is especially true if the long range (110 nm plus) high frequency surface wave (HFSW) radar is employed. Indeed, there are means readily available in most of those places to provide the close surveillance required. The establishment of the maritime IFF system as proposed here will also greatly facilitate this ship-tracking problem as well. In those areas currently without adequate radar surveillance a few well-placed aerostats such as those used in counter-drug smuggling operations should provide sufficient coverage. However, radar is not enough. Ship-based satellite communications transponders, serving an IFF function, are the key to solving the ship traffic-management system. More on this below, but it is necessary to discuss a few preliminary issues first.

Cargo and crew/passenger tracking

Monitoring the contents and tracking the location of containers is at the heart of shipping security. Regardless of the point of entry, whether land or sea, our greatest potential for security breaches and contraband comes from containers. The tracking of containers bound for the United States is a particularly important responsibility of the US Customs Service. The US Border Patrol, the Drug Enforcement Agency, and the Federal Bureau of Investigation, plus other law enforcement agencies, support Customs in this effort. These agencies also assist the Immigration and Naturalization Service, the agency responsible for vetting and tracking the people arriving in the United States via all means, including ships. This article deals primarily with ship tracking. However, due to the recognized critical nature of the need to track containers, the appendix to this article discusses some of the currently available technology in more detail.

The people-vetting and -tracking problem is also exceedingly difficult but will not be addressed here other than to note that the float plan, systems, databases, and procedures developed to track ships would assist the INS in its people-tracking efforts as well. Thus, though this proposal is focused on ship tracking for port security, it has much wider implications than just counterterrorism. It would also greatly assist in the war on drugs as well as help curb the illegal immigrant problem as well as help anti-environmental pollution operations.

The tracking of ships bound for the United States is a US Navy and US Coast Guard task, and this article deals primarily with the mission and tasks of the sea services: the USN/USCG team. Ship tracking is now undertaken only by exception when extraordinary circumstances warrant. This article proposes that ship tracking be done on a routine basis. It is clear that the technology is available, at a reasonable per unit price, to put a transponder in every container bound for the United States. However, the aggregate cost may well prove prohibitive. On the other hand, it is not cost prohibitive to put a transceiver or transponder on every ship and track it. This will be discussed in detail later in this paper. The payoffs against terrorist threats, as well as against drug and illegal immigrant smugglers and polluters could be substantial, far, far outweighing the cost.

Indeed, most of the civilian agencies named above already have some limited maritime surveillance capabilities to cope with those types of problems. As an example, the Customs Service has an excellent facility at March AFB, originally named the Air and

Marine Interdiction Coordination Center. (It is now known as the Air and Marine Operations Center). It is primarily focused on countering air smugglers, tracking all aircraft crossing any border in North America. The coverage of the radars displayed there, brought into the AMICC via live video feeds, actually extends far into, if not totally covering, most countries in North America and well into South America. They are also linked to multiple databases, including many law enforcement databases. They have a file of all flights flown by any aircraft in their database for the previous two years and can instantly pull data on any significant event or person associated with any aircraft they are tracking. Though currently AMICC makes only a minimal effort against marine smugglers due to manpower and equipment availability limitations, Customs would like to see that capability expanded. They clearly understand it needs to be done and given the resources they are ready to do it or help whoever does get the job. Their coordination of multiple reporting entities, and the software tools they have developed to assist them in their task, are especially instructive. The seven to eleven watch-standers routinely monitor all air traffic over the borders of North America, selecting an average of 2,900 tracks (out of tens of thousands) a day for special, more detailed examination. At any one moment there are about 5,000 aircraft airborne either over the US or in its immediate vicinity. The AMICC's experience in developing their system should prove very valuable in developing a maritime counterpart. The AMICC is also a major participant on the Domestic Events Network. The maritime surveillance organization proposed by this paper would also be a major participant on that network and closely interfaced with the AMICC.

Many of the technologies and procedures discussed here are pertinent to two or three of the tracking problems (people/cargo/vessels), and the research that led to this article is a great place to start the investigation of exactly what is needed in these areas; nevertheless, however closely related, the three tracking concerns remain clearly separate efforts, assigned to three separate governmental agencies. Tracking environmental polluters is yet another effort that would be assisted by the proposed system.

Framework of discussion

In the final analysis, the maritime homeland security/defense mission is a detect-access-act cycle. Most such cycles can be broken down several ways. The most famous model is the "OODA loop," which consists of the elements Observe, Orient, Decide and Act. Another widely held model is the Sensor to Shooter paradigm. The most recent manifestation of this cycle is the Find, Fix, Track, Target, Engage, Assess model. Though each of these models is useful, none fully describes what actually happens in a systems sense. We will use a slightly different model to describe the actions that need to be taken to build a vessel-tracking system and the interfaces needed to tie such a system into a decision-making apparatus both to effect timely action against potentially hostile threats and to apprehend others engaged in illegal activities.

This model, called "Warfare in the Fourth Dimension," was developed more than 20 years ago by the author while a research fellow at the Naval War College to describe and analyze the importance of "Time" to decisions in combat. It was first used to equate the battle for control of the electromagnetic spectrum with the battle for Time, the fourth dimension in physics. The model's components are the sensors (S), processors (P), fusion system (F), decision maker (DM), and action taker (AT), as well as the communications links that tie each of those components together. The paradigm closely mirrors what actually happens in all forms of combat, be it an infantryman fighting in close combat or a ballistic missile defense action on the edge of outer space.

Sensors detect signatures given off by potential targets and forward data to processors, which feed information to the fusion system. The fusion system provides knowledge to the decision maker. He, in turn, takes all other factors of the environment, including rules of engagement, force status, strategic situation, political alignments, and so on, into account, and develops as clear a tactical picture as possible and (hopefully) the wisdom that goes with it. From whatever wisdom developed, the decision maker issues orders to the action taker. The sensors detect the results, or lack thereof, and the cycle starts all over again.

In close ground combat, the eyes and ears (and hands and nose if the conflict is very, very close) are the primary sensors. The processors, fusion system, decision maker and action taker are all self-contained within the soldier, and the communications systems are the synapses in his brain. At the other extreme, on the edge of space, the sensors might be infrared or electronic intelligence satellites, linked to their processing centers on the ground by high-capacity data links that are, in turn, linked to the fusion system via military satellite communications (satcomms) or fiber optic cable. The fusion systems might, or might not, be collocated with the decision maker. Most likely the decision maker would be linked to the action taker(s) via a separate military satcom system. Battle damage assessment (BDA) uses the exact same systems; they are tasked to look for confirming phenomena, and then the S-P-F-D-A process starts all over again.

System of systems

The requirements for an enhanced tracking system are being widely discussed within the Navy and Coast Guard. The basic requirement for overall situation awareness is called Maritime Domain Awareness (MDA). It is analogous to the FAA's flight-following system. Numerous wargames and conferences, as well as my own research, show that various existing systems could be modified to provide the basic building blocks for a system to provide the necessary awareness; this would be the first step in building a NORMAD required to protect our coastal and maritime infrastructure. Stepping through each of the segments of the S-P-F-D-A model, I will describe how existing systems could be employed to build a system of systems to satisfy the requirements of building NORMAD.

Sense

The first step in this chain is to select the specific phenomena that can be detected by sensors and be processed by the rest of the cycle in a timely manner. This is the unique heart of the proposal.

Beyond the traditional sensors, such as radars and acoustic devices, there already exists a set of cooperative reporting systems, communications satellite-based identity and position reporting systems that are available over the ORBCOMM, ARGOS, or InMarSat communications satellite systems—each of which could be adapted for use as a primary sensor for Maritime Domain Awareness. GlobalStar and Iridium communications satellite systems may also be developing similar transceiving or transponding systems. Yet another company, Comtech Mobile DataComm, has developed a transceiver as a unit tracking system that could also be modified to be used in the envisioned system as well. I have named each of these companies as they are the only ones I have found after a significant search that appear to meet even the basic requirements. I would, indeed, like to know if there are other companies and systems that can provide the basic requirements for a maritime IFF system.

Each of these systems would need a common vessel identification scheme. Fortunately, one is readily available. In several of those systems the Maritime Mobile Service Identity (MMSI), assigned by the International Telecommunication Union (ITU), is used as the vessel/unit identifier. Discussions with developers of most of the other systems indicate their systems could be relatively easily modified to broadcast an MMSI as well.

The envisioned system would be the maritime equivalent of the aircraft system's IFF, with the MMSI being the identification code. In aviation systems IFF originally stood for the Identification Friend or Foe transponder interrogated by military radar systems, but now the term IFF is used to denote the primary electronic means of identification of radar tracks for both civilian and military uses. Radar is an integral part of the IFF system, interrogating the unit-based transponder and reading its response. However, to implement a fully useful ship tracking system such as would be required for NORMAD it will be necessary to track ships well out beyond land-based radar ranges, so I have looked to other technology, that of communications satellite transceivers and transponders, to fill the function of identification and location provided by the IFF system in aviation. As indicated above there are only five communications satellite systems that either now, or in the immediate future, can or will soon be able to meet the reporting requirements over the broad ocean area. Currently only two of these satellite systems, InMarSat and ORBCOMM, appear able to provide timely position reporting with oceanic coverage. Two other satcom systems are on the threshold of developing the needed capability, GlobalStar and Iridium. The fifth system, ARGOS, has an oceanic communicating and reporting capability but has significant time latencies built into its design. However, once a firm market and a known requirement exists, other satellite companies may well decide to provide the required services, either by adapting the capabilities of using existing satellite systems or by including oceanic capability in new ones.

Those readers seeking brief descriptions of the MMSI and the several satellite tracking systems suitable for maritime use can find them in the appendix. They can also find brief descriptions of two other pertinent electronic systems, the Automatic Identification System (AIS), the Digital Selective Calling (DSC) maritime radio, which also have significant roles to play in a maritime traffic tracking system.

Process

The signals containing the unit's identification and location would be broadcast via a transceiver or transponder onboard every ship desiring to enter the coastal waters of the United States. The signal would be received by one of several communications satellite systems now in orbit that have oceanic coverage, depending on which transceiver/transponder was installed.

Eventually, the effort could be expanded to include monitoring the Automatic Identification System (AIS) system as well, when AIS transponders are placed in orbit, as has been suggested or a method is found to route the AIS signal through one of the existing comsat systems. AIS is described in detail in the appendix but suffice to say here that AIS is an excellent, high fidelity collision avoidance and traffic management system just now coming into use. However, currently there are no satellites in orbit that can receive and process the signal. Preliminary discussions are underway regarding placing AIS transponders or receivers on spacecraft, but it is unclear when, or even if, this effort will come to fruition. This author strongly believes the effort to put AIS transponders in space or have the signal routed through the existing satcom systems should be accelerated, as the advantages to global shipping control would be significant. Manned or unmanned aircraft and aerostats could also be equipped to monitor AIS and used in a surveillance/patrol role, but I believe the system I am proposing here is a significantly less expensive alternative.

The gateway earth stations receiving the downlink from the satcomms with the ship reported position generated and transmitted over whatever system, be it InMarSat, AIS, or one of the other satcom systems, would forward the data to both the National Maritime Intelligence Center (NMIC) and MDA command and control centers that I called USCG Regional Reporting Centers (RRCs) in an earlier paper.

Since initially proposing the creation of RRCs, the Coast Guard has, with Office of Naval Intelligence (ONI) assistance, created two Maritime Intelligence Fusion Centers (MIFCs), one on each coast. Also, the Defense Information Systems Agency (DISA) is experimenting with a concept it calls the Area Security Operations Command and Control (ASOCC) which is a communications and software suite to link many of the organizations involved in homeland security. Whatever the site is eventually called or wherever it is located, be it at a place like the AMICC or at the NMIC or at the MIFCs, it would be linked to the satcom ground systems receiving the position reports, as well as to military and other government (USCG/USCS/DEA/USBP) forces in its area of responsibility. It would be the single entity responsible for tracking all vessels in its area or region and for orchestrating any tactical response required. It would be assisted by the NMIC's civilian merchant ship section, which would do the long-term trend analysis required.

Fuse

All-source intelligence fusion would primarily take place at the NMIC, but the MIFC, ASOCC and the existing fleet battle watch organizations maintained at each numbered fleet would assist in the effort. Coordination would be over SIPRNet (Secure

Internet Protocol Router Network) but because much of the data is not classified the unclassified web could also be used. Data from national collection means, over-the-horizon radars such as the ROTHR (Relocatable Over-the-Horizon Radar) and the HFSW (High Frequency Surface Wave) systems, sighting reports furnished by Navy and Coast Guard vessels and aircraft, human intelligence, and acoustic sensors would be melded together with the positions provided by self-reporting units over the satellite communications systems to determine the presence of non-reporting vessels or those with abnormal behavior or histories. This is not an insurmountable task as some believe. The AMICC investigates an average of 2,900 anomalous tracks daily. At the AMICC careful analysis and prompt information exchange with other governmental and private entities quickly clears the vast majority of unusual tracks but on a near-daily basis tracks are determined to be suspicious enough that the AMICC initiates an intercept of the suspect track by Customs aircraft. Similarly, organic fleet and USCG assets, both air and surface, could be dispatched to check on the few entities judged suspicious by the MIFC. Using their organic sensors, patrol units could quickly identify all vessels reporting their positions and thus could focus their attention on those not reporting or on vessels with reporting anomalies (such as a vessel using the MMSI of a ship known to have been in another part of the world very recently). Vessels not reporting or with reporting anomalies could be interdicted and interrogated to determine their status and intentions. The patrol units would be linked to the MIFC and its ASOCC via UHF satcom, and the MIFC would be linked to the vessel master file, probably at the NMIC. The vessel master file would contain everything known about the vessel and its owner(s) including the current float plan and all previous ones, any associated MMSIs, the ship's type, history of ownership, cargos, and crews, available from all other sources, plus any special notes, such as association with suspicious entities or activities.

Querying the vessel database using the MMSI would be much like a highway patrol officer's running a license plate check. A query to the Department of Motor Vehicles database can tell a patrol officer if a suspicious car should be pulled over. An MMSI check would provide the same benefit to our maritime forces. Establishing the MMSI as an IFF, or an "electronic license plate," would be of substantial benefit when a Navy or Coast Guard patrol unit has to "check out a suspicious unit and pull it over." It is realized the Posse Comitatus Act, signed into law in 1878, constrains some of the Navy's actions in this type of tactical situation, legal means can certainly be found to halt a suspicious ship on the high seas. The USA Patriot Act of 2002 allows military platforms to collect intelligence on civilian entities in the manner described here.

Establishing the MMSI as electronic license plates and developing the means to track them are small but important steps in the chain. They fill a substantial void in the maritime defenses of our nation. It is true that getting all units approaching the coast of the US and its territories to broadcast their MMSI would require a cooperative effort. However, this government can enact regulations that would require all vessels desiring to enter US ports to commence broadcasting their MMSI over one of the approved long-haul communications systems either when within a specified distance (1000 NM?) of the coast or when they are within 96 hours of entering port. Legally operating entities will enjoy greater safety with the additional track following. Any ship not filing a float plan or not broadcasting via one of the approved systems would be immediately obvious to patrolling units and could be stopped for investigation. Those not broadcasting their identification, location, course, and speed would be subject to interception and inspection, causing significant delays in entering port, if indeed, they were allowed to enter port at all. Thus, the impetus to all vessels to broadcast their identification (the MMSI) as well as location, and file a float plan, would be substantial. Delay costs money—more money than acquiring communicating systems that most ships already have, or should have, for safety of life at sea (SOLAS) issues, in the first place.

The processing system outlined above is an expansion of capabilities already in place at the Joint Inter Agency Task Force (JIATF) facilities on both the east and west coasts of the United

States and at the AMICC. In addition to the need for a universal, mandated maritime position reporting and float plan notification system such as proposed here, sufficient people will have to be employed or enlisted in order to maintain a full maritime watch. Fortunately, software tools in use at the AMICC and at other government agencies such as the National Security Agency (NSA) and the National Reconnaissance Office (NRO) have shown that the manning requirements can be quite small.

New-generation display and decision (D&D) technology, such as that contained in the Anti Air Defense Commander (AADC) system developed at Johns Hopkins University's Applied Physics Laboratory (JHU/APL) with easily understood symbology and embedded reasoning and data manipulation capabilities, and which are now being deployed on Navy command ships and cruisers, could be used to help the MIFC gain and maintain situational awareness. This AADC-like D&D system would be the focus of the data fusion efforts. Software tools called smart agents (SA) would be used to sort the huge amount of data flowing into the MIFC and ASOCC. SA tools are discussed in more detail below. The envisioned D&D system could also be a significant instrument for managing communications links into and out of the ASOCC.

Decide

A correct decision requires a sufficient quality and quantity of information in sufficient time to fuse it to develop knowledge. The timeliness requirement dictates that the decision maker knows when he/she has the required minimum sufficient amount of information. That knowledge is combined with additional information from all other sources and includes all aspects of the problem at hand, such as status of own forces and rules of engagement, as well as the political, strategic, operational, and tactical situations to develop wisdom and to issue the appropriate orders. This is by no means a trivial task; indeed, integrating vast amounts of data from heterogeneous sources is exceedingly difficult. The complexity of finding and integrating information from disparate sources is daunting for the human mind; fortunately, however, several software tools are now available to help the decision maker in this task. One of these is the Architecture for Distributed Information Access (ADINA) tool developed at JHU/APL. ADINA is an agent-based architecture for seamless access to and aggregation of heterogeneous information sources.

Procedures and smart agent tools such as ADINA and the CoABS grid, described in the appendix, would be in place at the MIFCs and its associated ASOCC both to fuse the data coming in from all sources, including the crucial MMSI data, and to formulate decisions to be made and courses of action to be followed, in close coordination with the US military command structure in the immediate area. The D&D system for the decision maker, as envisioned above, could have its own CoABS grid to assist in developing Wisdom by considering all relevant situational data. The smart agents would also assist in issuing the appropriate orders to the appropriate units—orders that are correct for the tactical, operational, strategic, and political situation.

Act

Once the decision is made to interdict a specific vessel—most probably because it is non-reporting, or because intelligence indicates it may be a potential threat or is engaged in illegal activity—the Navy, in previously arranged consultation with the USCG, would assign an on-scene commander as officer in tactical control (OTC). There are rules of engagement in place that clearly spell out which organization is the lead federal agent in all anticipated cases. Military forces, including surface forces and air elements of both the USCG and the USN, as well as USAF units would be assigned to take appropriate action. As speed of response could be crucial in some tactical situations, those units should include such USAF and USN regular and reserve forces as A-10s and P-3s, equipped and trained for anti-shipping attack with appropriate weapons. Those weapons should include optically guided missiles such as Penguin and Hellfire to allow disabling fire to be focused on the bridge and rudders of rogue ships attempting to enter port with clearly hostile intentions. In extremis, such as might be the case of a ship known or strongly suspected of carrying weapons of mass destruction, larger weapons, such as Harpoon, must be readily available to sink that rogue ship. If more time is available and forces are in position, then surface units could conduct the interdiction. Helicopter insertion of special forces and specially trained units are also possibilities.

Navy and Coast Guard vessels and aircraft routinely operate off our shores. Under this proposal, they would both have the responsibility to report all surface vessels in their area as well as act as "first responders" in any interdiction. Under the procedures outlined here the reports from these units would be fed into the vessel database maintained at the NMIC and matches made automatically to the pertinent float plan. Non-reporting or suspicious vessels would be marked for follow-up investigation and possible interdiction.

All units of the US government assigned to surveillance and interdiction roles should also be equipped with DSC and at least the ability to monitor AIS, if not be fully partici-pating units in AIS. Both of those radio systems allow another entity to interrogate a specific vessel and determine its MMSI, and thus its identity if the MMSI database is available as this paper suggests. As indicated above, it is analogous to a highway patrolman being able to read a license plate and query a motor vehicle database. Having the responding units AIS and DSC equipped would allow them to quickly determine which vessels in their immediate area (line of sight) are suspicious and need to be investigated further and possibly stopped.

Concept implementation requirements

This proposal envisions the establishment of US regulations, starting with a Notice to Mariners (NOTMAR) requiring all vessels operating in the belt 50 to 1000 nautical miles offshore and desiring to enter our territorial waters to broadcast their identification and location at set intervals over any one of the approved basic systems. If a vessel is coming from beyond 96 hours sailing time away it needs to have been broadcasting its location since it was 96 hours out. If it is coming from a nearer location the vessel needs to have the system on as soon as it is underway.

A final word on available technology. The International Maritime Organization (IMO) already requires units above 300 gross tons (gt) to be INMARSAT-C equipped as part of the Global Maritime Distress and Safety System (GMDSS) in accordance with the SOLAS convention. These same vessels will also be required to have the more expensive and more technically sophisticated AIS by 2008. Under this proposal, they can report via any approved system they select. Once AIS is capable of being monitored from beyond line of sight (via space systems?) it may well become the system of specification. Until then, they can use either their existing INMARSAT-C system or one of the other significantly less-capable and less-expensive systems either now available or coming on-line soon. This proposal would require all vessels including those under 300 gt operating more than 50 miles offshore to purchase and operate a satellite communications reporting system based on systems such as ORBCOMM or INMARSAT-C (or GlobalStar, when it comes on-line in the near future or Iridium whenever it is available). Another system to be considered is the Comtech Mobile DataComm transceiver/transponder. It was designed for Army use but it uses any L-band satellite, and InMarSat is an L-band system. Some cost sharing in the form of tax relief could be worked out to assist American vessel owners with the installation of at least the basic system. For foreign owners, the cost of entering US waters will increase, but not by an unbearable amount. Operational tests would need to be run on each of these systems to ensure they do operate in a sufficiently timely manner and that their data stream could be made compatible with a national reporting standard. The task is clearly doable from a technology view.

Summary

It would be very much to the benefit of US security, maritime and otherwise, if the system and legal requirements outlined above were enacted in the immediate future for vessels bound for the United States and areas under its protection. This article is provided as a point of departure as we build the maritime portion of Homeland Security mission capabilities package. If more capable systems will soon be available, or a more beneficial alignment of existing systems can be made, so much the better. As the many people that have seen this article change and grow over the past year will attest, it is meant only to further the discussion and provide an informed point of departure; it is intended as a "thought-starter," not as the final word. The author is profoundly grateful to the many people who have made significant inputs to the furthering of this concept by providing additional data-points and suggesting additional emphasis on pertinent aspects of the concept.

Possible NOTMAR wording

"Be advised: All vessels intending to enter or transit the territorial waters of the US or its protectorates (Guam, Puerto Rico, Virgin Islands, etc.) must file a float plan with the US Coast Guard, prior to arriving within 96 hours of the coast of the US or its protectorates. If the point of departure is within 96 hours sailing time the float plan must be filed prior to leaving port. The float plan will include:

1. The names and nationalities of all persons onboard.

2. List of all MMSIs to be used on the voyage.

3. Description of any and all cargo.

4. Point of last departure.

5. Destination.

6. Estimated time of arrival

7. Estimated time and location of arrival at a point 50 nm form the coast of the U. S. or its protectorates.

Additionally, all vessels must also have one of the following systems on and transmitting its identification (MMSI) and location. It must report the vessel's position and MMSI (ID) not less than once an hour when within 500 to 1,000 nm from the U. S. or its protectorates. When within 500 miles of the United States or its protectorates and planning on entering US territorial waters the vessel must broadcast its identification and position four times an hour. Vessels not complying with this directive will be subject to interception and detention for a minimum of 24 hours at the limits of US territorial waters."

Current Systems

InMarSat-C, ORBCOMM

Systems under review

AIS via a satcom system(s) oceanic coverage, GlobalStar, Iridium, Comtech Mobile Data Comm (InMarSat).

Appendix One Expanded:
Maritime Traffic Systems
for Maritime Domain Awareness

Introduction

Following is a description of several of the systems and concepts discussed above. It is included with the paper, as it is believed a basic understanding of these systems and concepts will significantly assist in understanding the issues involved and the concept proposed. Several significant maritime communications systems are available that, by providing an offshore tracking capability, could serve as Maritime Domain Awareness (MDA)'s situational awareness backbone. Those systems are the Automated Identification System (AIS), the Digital Selective Calling (DSC) capability, as well as the several communications satellite systems which are capable of operating over the ocean with automated location reporting capability (such as InMarSat and ORBCOMM). An integral part of the concept is the need for a universal identification scheme. That role is filled by the Maritime Mobile Service Identity (MMSI), an electronic license plate for ship-borne radios. All these systems, plus a more in-depth look at container tracking, the Area Security C2 Center (ASOCC), multi-level security, smart agents, and HF/MF ship communications systems are also described in this appendix.

This discussion is not an endorsement of any system or set of systems; rather, it is a description of the technology that is in place now or will be in place very shortly. It is entirely possible that, given the rapid advances in technology, a more capable system could soon be developed. The ORBCOMM Comsat system is discussed as representative of the several different commercial satcom systems. Further study needs to be undertaken to determine exactly which satcom system(s) is most cost effective in this role; several of the systems may have, or perhaps could be easily modified to have, the basic automated position-reporting capability required. The basic reporting requirements need to be developed, documented, and published. Then all possible systems need to be operationally tested to assure they meet those requirements. Those requirements include the obvious ones of being able to report the required information in a timely manner, but also include the ability to not be spoofed or moved from the assigned ship without it being noted.

Table of contents for Appendix One Expanded:

1. Maritime Mobile Service Identity (MMSI)

The MMSI is a nine-digit entity designed to be "transmitted over a radio path in order to uniquely identify ship stations, ship earth stations, coast stations, coast earth stations, and group calls. These identities are formed in such a way that the identity or part thereof can be used by telephone and telex subscribers connected to the general telecommunications network principally to call ships automatically." In the US, the National Telecommunications and Information Administration assigns federal MMSIs. The Federal Communications Commission assigns Nonfederal MMSIs, normally as part of the ship station license application; the MMSI is assigned whenever a vessel purchases either an AIS system or a DSC maritime radio. The MMSI could also be tied to a physical description of the vessel, including its color, length, and type as well as information on its registration, cargo, crew, and ownership. This data would reside at the National Maritime Intelligence Center (NMIC) for query by all interested parties. The Appendix contains more detailed descriptions of the AIS, DSC, and ORBCOMM systems. Additional information can be obtained at the USCG Navigation Center web site at https://www.navcen.uscg.gov/gmdss-compliance-requirements. Inquiries can be directed to nisws@navcen.uscg.mil.

2. Automatic Identification System (AIS)

The first and most capable system to be discussed is the AIS. It is an International Maritime Organization (IMO)-approved system developed at least in part by the USCG. Its primary component is a small broadcast transceiver, broadcasting at 160 MHz in Time Division Multiple Access (TDMA) format. As a VHF signal, AIS propagates only to line of sight (LOS), so it can only be used when within LOS of another transceiver, either on another mobile user or on a shore site. The IMO has agreed to require that one of these transceivers be active on every commercial ship greater than 300 gross tons by 2008.

The system provides identification (MMSI), location, course, speed and a time stamp; which allows the AIS terminal to calculate range and bearing from all other units in range. This capability allows the crew to quickly calculate the closest point of approach (CPA) to all other AIS units in range, and thus is an excellent collision avoidance tool. AIS also provides vessel tracking information enabling traffic control for the USCG Captain of the Port (COTP) and his foreign counterparts worldwide. The system's collision avoidance capabilities alone are sufficiently beneficial that most merchant ships leave it on at all times, even on the high seas, to warn the ship's crew when potential collision situations develop. Ships are being equipped now, and the USCG has a full-up system deployed to San Francisco Bay as a test concept. Several other ports have some element of the system installed as well.

The USCG R&D Center in New London, Conn. is interested in seeing if this system could be expanded to include a capability to monitor its transmissions from space to expand the USCG's ability to track ships well beyond the current limitations LOS imposes. The IMO is interested in this capability for worldwide application. Being able to monitor AIS from space would allow tracking vessels at all times that the system was operating. The space segment could be used from the time a vessel leaves LOS of the AIS system at point of departure until it is within LOS of shore-based systems at the point of arrival. This would provide tracking out well beyond the 1500 NM from the coast envisioned as a requirement for MDA. The idea of developing the capability to monitor AIS from space has become a much higher priority since September 11, and senior Satcomm engineers at both JHU/APL and ORBCOMM do not see any substantial technical challenges with the concept. ORBCOMM is investigating the idea of including a receiver capable of receiving the AIS signal in the next iteration of its satellites, scheduled for launch in 2005.

The space segment for monitoring AIS is important because it could be used to cue other assets, allowing them to separate "wheat from chaff" much more quickly and focus on vessels that need to be investigated more closely. Space systems would provide continuous tracking and facilitate identifying potential problem vessels much further from shore, allowing more time to determine the correct course of action.

3. Digital Selective Calling (DSC)

The second system to be examined is the DSC capability of newer maritime VHF radios. It, too, has finished basic development and is currently being sold all over the world. DSC is preset as Channel 70 of these newer maritime VHF radios. It is configured as a search and rescue (SAR) system with additional selective calling capabilities, but it does not have the TDMA or any other modulation scheme that would permit multiple users to share the same frequency as AIS. It quickly becomes self-jamming when more than a couple of units are active at the same time in the same area, so it is not recommended for use as a wide area surveillance system. In local areas, however, it is an excellent identification and tracking tool, as it also employs the MMSI and can be coupled to a global positioning system (GPS) or long-range navigation (LORAN) to automatically provide the unit's location whenever it is transmitting. As the location can be manually inserted, this system can also be used to send false and misleading position reports. In contrast, GPS is an integral part of AIS and ORBCOMM and is thus much harder to spoof.

4. Commercial Satellite Communications Systems

The third set of systems under consideration is the various commercial communication systems that routinely operate over water. Those systems include the several manifestations of InMarSat, a geo-synchronous satellite communications system, the Argos system that operates in a highly elliptical orbit, and the low earth orbiting ORBCOMM, GlobalStar and Iridium communications satellite systems.

We will describe the ORBCOMM systems as representative of the capabilities available to the low earth orbiting communications satellite systems. Likewise, we will describe the Secure Asset Reporting System (SARS) as representative of the display and decision systems now on the market.

As an on- and off-shore tracking device, the ORBCOMM commercial communications satellite system can receive short, formatted messages from mobile units to provide identification, position reporting, and tracking data at user-predetermined intervals. The system consists of approximately 35 small communications satellites in low earth orbit (500 NM) and earth gateway stations in many coastal areas, both in the US and abroad. The individual subscriber unit is a small transmitter with an internal GPS receiver. This unit could be used to track vessels but would require placing the transmitter on each vessel. The cost of the basic, minimal-capability shipboard system is less than $500, plus the cost of a simple VHF antenna and attachment to the ship's power. Other units with additional capabilities, such as automatic telemetry broadcast, are available at additional cost. In SARS crew and passenger lists, as well as cargo manifests and the float plan, can be, and often are, sent ahead via e-mail over landline before the vessel sails. These documents are stored in the SARS database, linked to that specific vessel. Changes to those documents in SARS are made via short messages over the ORBCOMM system. Both the original data and its changes, with the date the changes were made, are retained for two years in SARS. The elegance of this is that there is no need to send large messages over ORBCOMM, a narrow bandwidth system.

Should anyone in a company who is using SARS (and has access to his company's SARS password) need to know who or what is on the ship, or was on the ship in any previous voyage, all they have to do is pull up that trip record and select the record for review. As indicated above, this really saves bandwidth, yet all the data is in an easily retrieved database. Several other communications satellite systems could probably fulfill the role described above. Additionally, there is reported to be something on the order of 30 display and decision systems such as SARS now on the market. If a reasonable standard is set the competition between the several systems will be on price versus performance, and will not be dominated by any one system, as maritime satellite communications are today.

The INMARSAT systems, especially InMarSat-C and Mini-M, and the ARGOS systems were also considered, but the ORBCOMM system's simplicity and its ability to rapidly forward position reports, coupled with its low cost to both acquire and operate, dictated its selection. However, to accommodate ships already equipped with InMarSat-C (which has substantial additional communications capability beyond position reporting), provision could be made to accept either the InMarSat-C or the ORBCOMM system, or any other forthcoming comsat system, in the track-reporting schema to the envisioned ASOCC. The ARGOS system appears to be too time-late to be used as an MDA tool. It would not be a technical challenge to develop a software translator that converts the messages received from either system into the Navy's rainform gold reporting format, or a format that could be input directly into the military's Global Command and

Control System (GCCS) and its Navy component GCCS-Maritime (GCCS-M). ORBCOMM has reported similar data for various experiments.

Because the ORBCOMM system does not necessarily use the MMSI as the vessel identifier, it may also be reasonable to require all non-AIS-equipped ships to also be equipped with DSC, which does use the MMSI. Indeed, it is a good idea to direct all US vessels capable of operation beyond twelve miles at sea to become equipped with DSC, for Safety of Life at Sea (SOLAS) considerations as well as to further ease vessel tracking and identification in a localized area.

5. Container Tracking Systems

One other aspect of the various Satcom transponder-tracking systems is their ability—already in widespread use—to track high-value/high-interest containers from point of origin to final destination. If the United States decided to put transponders on every US-bound container worldwide (500,000 units?), the cost would be about $170 per container, plus installation. (Installation costs vary depending on the type of container—dry cargo, etc.—and time of installation—at manufacture or retrofitted.)

Significantly more capable transponders, which can also be configured to tell if and when, and for how long, the container has been opened or gone out of a specified temperature range (important for fruits, vegetables, meats, etc.), would cost in the $300 range. These transponders can be programmed to automatically switch from reporting over the ORBCOMM system to GSM cellular telephone systems, the system in use throughout the world and now one of the three primary systems in the United States.

GSM is also useful for vessel tracking because it provides a high rate of position reporting should an agency desire to track a container or vessel in port at a higher rate than is feasible with the satellites. Once the ship carrying the container is within LOS of a GSM cell phone tower on the coast—approximately seven miles out in some cases, less in others—GSM also provides redundancy and the capability for an agency to contact the bridge on a secure voice channel via cellular phone.

The following is an e-mail I received from Mr. Jim Kross, general manager of ESL, LLC, a value-added reseller (VAR) for ORBCOMM. His company markets ORBCOMM services and he is teamed with a company that makes the mobile transceivers as well as one that has developed the tracking and display software, Secure Asset Reporting System (SARS).

> "It is technically feasible for the system I described, in its current configuration, to support all 30,000 containers entering the US daily, as well as all the containers entering Canada and Mexico, and track each container to its final destination. In the US, the aggregate number of units is approximately 1,000,000 but our studies show that under present patterns of use only about 1/3 would report daily (incoming and outgoing units stacked in port will not transmit unless there is an alarm).

> "Furthermore, our dual-mode transceivers utilize the GSM network when available and as needed to supplement the ORBCOMM constellation, extending the current system wide capacity in North America to many times the above numbers. This also reduces the latency of reporting in dense urban areas, at the point of origin as well as in the US, and anywhere where space-based systems are blocked.

> "I should stress that the wireless segments of this system, which rely entirely on commercially available components, were designed from the inception to send highly compressed short messages. For example, a time stamped position report is only 6-bytes of data in a 15-byte packet. The system relies on the Internet and not the wireless segments to provide large documents such as crew lists, ships manifests and customs documents, etc.

"Also, most of the intelligence required to provide the functionality described is contained in the transceiver and not on the ground segments, further reducing communications overhead and improving overall redundancy and availability. The on-board intelligence is used to adjust transmission rates, monitor status, and send alarms, all of which reduces the communications overhead and operational cost.

"The current cost of the described equipment installed in containers in unit volumes of hundreds is acceptable only for high value/interest shipments. In unit volumes of tens-of-thousands the costs drop dramatically. Transporters currently use our system for many commercial reasons, among them improving demurrage billing and thereby reducing the length of time a container [goes] without earning revenue and the number of containers in circulation. Some companies estimate that they can reduce container inventories up to 15%. The other advanced features available to shippers (separate from the security features) are designed to reduce operational costs through automation, consolidation, and centralization of various manual processes. If adoption of this hardware/service standard would allow shippers and transporters to develop a trust relationship with US monitoring agencies and thereby achieve the benefit of reduced entry time into the US, Canada, and Mexico, the unit cost of the hardware could be further justified.

"Ultimately, there must be a convergence of commercial and security interests to make this level of sophisticated vessel and container management available on a wide scale and achieve the cost reductions and level of security desired by all parties.

"1) All satellite transceivers have a secure numerical identifier and Internet address that can be associated with a Mobile Maritime Service Identifier (MMSI) at the time of provisioning within the satellite service providers secure databases. For example:

"Unit Serial # / Internet address / MMSI Vessel / name / Associated data

"165STREL / A20@yahoo.net / 12345678 / Forrester / Other data

"All the above data is maintained in a secure database accessible exclusively by the satellite service provider. Security procedures and audit standards for these databases are highly developed. Provisioning the MMSI identifier for vessels equipped with ORBCOMM satellite communicators entering US waters would be a function conducted exclusively within the United States at a single facility in Virginia. Once provisioned, MMSI data could be made available to selected law enforcement agencies worldwide.

"All the associated data, such as cargo manifests, transport routing, crew lists, etc., are maintained in trust by agreement with the exporter, importer, shipper, foreign and domestic customs services, and other government authorities. With appropriate electronic customs seals on cargo containers every shipment can be continuously tracked from the point of origin anywhere in the world to its final destination in the US, US Customs and other civilian and military authorities would be able to view the manifests as soon as the containers are sealed, and the shipper posts the data electronically. Although currently the form and substance of the electronic postings vary from country to country, sufficient data is available to [form] the basis of the system you propose. Future standardization would be desirable.

"Any vessel with an ORBCOMM satellite communicator that will automatically process incoming GPS data and transmit periodic position reports can be MMSI compliant. Furthermore, this two-way system can be selectively interrogated, forcing the on-board communicator to immediately send a position report, change the frequency of reporting interval, and query attached sensors. These functions are controlled within a secure environment that cannot be modified on-board.

"The interval for automatic position reporting may be preprogrammed in the non-volatile firmware of the satellite communicator and can be set to report anywhere from once a day to once per minute. For instance, a reporting interval for vessels at sea might be 4 times per day, then automatically increase to once per hour when the vessel crosses a specified threshold such as 500 miles from port. At the 12-mile limit the interval might increase to 4 times per hour and within the boundaries of a port increase again to once every 10 minutes or less. Using other preprogrammed features, the unit can report any time it is rebooted, when the vessel begins movement, stops movement, or slows below a specified threshold, etc. A dual-mode device containing both satellite and ground-based transceivers provides communications redundancy as the vessel enters US waters.

"Satellite communicators placed inside cargo containers can be programmed to send a security report when the container is sealed at the point of origin and then track the unit while in transit to the point where it is loaded on a vessel. Whenever the container is removed from the vessel it will again report its position. If at any time the container is opened after it is initially sealed, then an alarm is transmitted automatically. Sensors inside the satellite communicator automatically report on conditions within and around the containers. This cradle-to-grave monitoring is economically practical with existing hardware. The only question today is whether to retrofit existing containers or build the equipment into new units only."

6. Multilevel Security

Mr. Kross also shared his insights on what depth of coordination, assisted by existing technology, will be required to allow successful tracking of the containers, once they are within the United States, across several law enforcement agencies with different security classification systems. This is a very real problem that has plagued the military for years and is now greatly complicating HLS efforts due to the addition of LEA classification systems. Mr. Kross comments:

"Backoffice associations, the trust relationships required by civil and military authorities, the on-board hardware and software are all available and in commercial use today. The infrastructure allowing access to and sharing the desired information, however, is not as readily available. What appears to be missing is a means to provide the appropriate degree of access to users with differing authorizations.

"Good models already exist, such as Pennsylvania's Web-enabled statewide criminal Justice Network (JNET) (recently demonstrated to Admiral James L. Loy, Ret. USCG Commandant, head of the new Transportation Security Administration), which is the result of an initiative undertaken by former Governor Tom Ridge. JNET provides a virtual single system based on open Internet technologies with standards that link information from diverse, seemingly incompatible systems of 16 different criminal justice agencies. The system enables agencies to share information but does not affect independent operating environments. As required by certain confidentiality statutes, each agency can determine the extent to which the others have access to its data. JNET is a secure extranet providing a secure publish and subscribe architecture featuring encryption and digital user/server authentication certificates."

7. MF/HF Systems

It has been estimated that up to 80 percent of all commercial ships still employ radio teletype (RTTY) such as Clover-II and Clover-4; Dataplex (GW-Pactor); Pactor-II; or Pactor III as their primary means of communications. Additionally, between ten and 15 percent of all commercial ships still rely on either constant wave (CW) manual Morse or single side band (SSB) voice as a major communication system. However, the satellite transponders are now inexpensive enough that I believe it is time to require all ships entering our territorial waters to be equipped with them, just as we require all aircraft except the very smallest to be equipped with an IFF.

Given that AIS and DSC also broadcast position and identification information, if only via LOS communications, it still seems greatly beneficial for equipping US maritime patrol forces with the ability to monitor both the AIS and DSC systems, as well as to be tied into an ASOCC via UHF satellite communications systems. The addition of AIS and DSC systems to US patrol forces would provide them with the ability to ID any unit quickly and accurately in its patrol area broadcasting these internationally mandated systems. UHF Satcom capability will allow them to query the shore-based databases and have rapid access to analytical capabilities not normally resident in patrol forces.

8. Area Security Operations Command and Control (ASOCC)

The ASOCC is a DISA ACTD to build a command and control element for homeland security. Conceptually, it contains all the communications capabilities envisioned in this paper. It would be the link to both the NMIC and the Combatant Commanders, as well as to the MIFCs. The ASOCC would link NMIC and the MIFC to the US law enforcement agencies (LEAs) and their intelligence efforts. The major US LEAs include the FBI, the USCS, the DEA, the USBP, and the INS as well as the USCG.

DISA is also developing an overseas counterpart to the ASOCC, the Coalition Rear Area Security Operation Command and Control (CRASOC), to tie US overseas rear area force protection assets into host nation force protection assets for mutual support. Its test bed is in Japan at the headquarters for the Commander US Forces Japan (CUSF-J). In fact, the CRASOC concept predates the ASOCC development. Once 9/11 happened it was realized that the same capability, as was proposed in the earliest versions of this article, did not exist and the ASOCC concept was developed. The bottom line is that the engineering design model (EDM) for both the ASOCC and the CRASOC are in use today and further development is underway. The proposed maritime HLS center would also be linked to both the US military command authority in its area and the NMIC via wideband communications channels using the ASOC. The ASOCC would also provide wideband communications linkage to civilian satellite communications systems as well as military command and control systems including the NMIC. This would provide access to the databases of such organizations as Interpol that NMIC is already linked to and may be in the future, including potentially various commercial databases, such as that maintained by Lloyds. These tie-ins are especially crucial for tracking and identifying cargo and crew, but they could also have a major impact on efforts to identify and track suspicious and potentially hostile vessels. The ASOCC would assist local fusion points for all information in a region and provide linkage to the USCG Maritime Security Squadrons and USCG Captain(s) of the Port (COPT) and Portmaster(s) of a region, as well as the MIFC in the area.

9. Smart Agents (SA)

Software tools called Smart Agents (SA) are now being experimented with in several war-fighting roles. One of the SA efforts getting a good deal of attention is the Defense Advanced Research Program Agency's (DARPA) Control of Agent Based Systems (CoABS). It arrays Smart Agents in grids assigned to perform specific warfare-associated information manipulation functions. CoABS grids could assist in processing data from the sensors, automatically routing items to databases while passing tasking refinements back to the sensors and passing on to the fusion system the information elements derived from its processing of the sensor data. The fusion system could, in turn, have its own set of CoABS grids to assist in the fusion function, passing necessary tasking refinements back up to the processors and sensors while routing correlated or collated data on to the decision maker's D&D system.

10. Summary

A two-tiered tracking system could be quickly emplaced, combining AIS and satellite transceiver systems such as ORBCOMM, or some other communications satellite system with similar capabilities (such as those included in some of the InMarSat systems). The tracking output from both systems, AIS and the communications satellite systems, could be forwarded via AMICC to the MIFC for tracking in much the same way as an FAA regional center tracks aircraft. Regulations could also be enacted to require all ocean-going vessels of all tonnages to have one system or the other. Obviously once AIS has the ability to have its signal broadcast to a satellite and received and processed ashore, it is the preferred system, but given its cost, at least $12,000 to $16,000, smaller craft may wish to employ one of the less costly (<$200?) satellite transceiver systems. Indeed, indications are that the widespread use of transceiver systems would drive the price down considerably.

Appendix Two
A Systems Engineering Approach
to Building a Maritime NORAD—
The Maritime Traffic Tracking System

Author's Note: Following is an ongoing revision of my thinking about maritime security provided to my list of some 1,100 contacts. that i developed as a direct result of my task in 2001.

I have participated in a few dozen wargames, seminars, symposiums, conferences, and other formal meetings in the that have dealt with "objectives definition" for maritime surveillance. This paper is an attempt to move the process past that level and commence building a solution. The US Coast Guard has established a combined, multi-agency Maritime Surveillance Working Group that is also working toward building a solution and, at the direction of the Commandant USCG, is developing a North American maritime surveillance plan (NAMSP). This paper is meant to be adjunct to that plan. It addresses a core capability of the NAMSP, that of a maritime border surveillance system.

This paper proposes a structured analytical approach to the multi-departmental, complex problem of building an effective maritime border surveillance system by using a series of WALEXs and 'LOEs (Limited Objective Experiments). WALEX is a systems engineering process/tool developed at Johns Hopkins University's Applied Physics Laboratory. It gets its name from the room in which the process was developed, the Warfare Analysis Laboratory. Thus, an exercise in the WAL is a WALEX. Coupled with the WALEXs would be a series of limited objective experiments (LOEs) using the US Customs Service Air and Marine Interdiction Coordination Center (AMICC) and the two new USCG Maritime Intelligence Fusion Centers (MIFCs) and the NMIC. This dual track of WALEXs and LOEs would allow the maritime surveillance community to "build a little, test a little" as it develops the complete system. This is necessary to gain a full understanding of the utility of greatly expanded maritime border surveillance and its impact over the complete operational spectrum both within each affected department as well as across departmental boundaries. This deliberate approach would allow for the most efficient planning for, and implementation of, the seamless integration of the multitude of requirements and assets involved, while quickly giving us a vitally needed maritime border surveillance system.

First, in the spirit of full disclosure, let it be clear that the author has spent many years in wargaming, including several working with and in the WAL and thus could be accused of bias toward its capabilities. If there is a better technical wargaming facility and, more importantly, trained technical war game developers and facilitators, then that is the one that should be used. This author is not aware of a better one.

Also, I want to define a couple of the semi-unique terms I have developed since 9/11 to assist in the debate on how to secure our maritime assets:

Maritime Traffic Tracking System (MTTS) – A combination of InMarSat-C polling as used on major sea-going vessels, a separate system of satcom transponders configured to provide IFF functions (Maritime IFF) for smaller ships, and AIS fused together with all source intelligence whose analysis is assisted by software smart agents to provide Maritime Domain Awareness (MDA) off our shores. MTTS is the information backbone for the proposed maritime NORAD.

Maritime NORAD – MTTS plus the command and control elements as well as the patrol and interdiction units tasked and trained to intercept, identify, and interdict maritime threats detected by the MTTS. In the parlance of my previous papers on this subject which described the Sense-Process-Fuse-Decide-Act process, MTTS provides the Sense-Process-Fuse portion of the loop while the maritime NORAD would be configured to complete the loop by providing the Decide-Act portions of the system/process. Now let us describe the process by which such a system could be built.

The WALEX process, first used to assess the impact of various technologies on all aspects of naval air defense, could be used to examine the maritime NORAD concept and develop a full understanding of the opportunities generated for MDA by MTTS and a maritime NORAD system. It could also be used to assist in formulating tactics, techniques, and procedures (TTPs) to both optimize this capability within our own forces and counter the threat. The WALEX process proposed as the principal tool for the systems engineering examination of this problem was initially designed to allow for the structured examination of complex problems with significant technical aspects, specifically, the myriad challenges of naval air defense. However, the WALEX process has expanded over the past 20 plus years and is now routinely used to gain a much fuller understanding of the requirements and impacts across all aspects of doctrine, organization, training, material, leadership, personnel, and facilities (DOTMLPF). WALEXs are also now routinely used to gain insights in many warfare areas, as well as numerous other less highly technical scenarios such as disaster relief, updating highways, and biomedical research. It is also used to assist in formulating tactics, techniques, and procedures (TTPs) to both optimize a capability within our own forces as well as to develop counters to a threat. In such a manner a WALEX could be used to bring a better focus on the threat to critical elements of our maritime infrastructure posed by potential opponents as well as to assist in the optimization of the capabilities of our own resources in a counter-terrorism role and other governmental maritime roles such as counter-narcotics operations, environmental protection, off-shore fisheries surveillance, illegal immigrant smuggling prevention, as well as safety of life at sea operations such as search and rescue, and medical emergencies at sea. LOEs could be used to test the results in a real-world environment and to train the various elements of the detection-to-analysis-to-decision-to-action chain.

Planning for an expanded border surveillance system had been ongoing for some time prior to 9/11. These plans were aimed at counter-narcotics and counter illegal immigrant infiltration operations. The recently recognized threat of terrorism has prompted the establishment of a Department of Homeland Security (DHS) and within that department, a Directorate of Border and Transportation Security (DBTS) that integrates border management agencies.1 The Homeland Security Act of 2002 charges the DBTS with securing the borders, territorial waters, terminals, waterways and air, land and sea transportation systems of the United States and managing the nation's ports of entry. The same act also establishes the Bureau of Border Security, headed by an Assistant secretary in the DBTS. It also establishes a Director of Shared Services.

The focuses of the Department of Defense and the US Coast Guard have undergone similar changes but how all these organizations are going to work together to protect our ports has not been fully established. Although good work is being done by several organizations in this area, it is obvious that the optimum force mix has not been established yet and much more needs to be done in this area. Indeed, there are senior naval officers that assert the mission of the Navy in homeland defense is to be carried out as far forward as possible, by attacking the enemies of America on their home ground, before they ever get the chance to get across the ocean that has protected America for nearly 200 years. However, one of the lessons of 9/11 is that those oceans are not the stalwart ramparts they once were. A determined enemy can, by using stealth, cross those heretofore stout bulwarks and do us grievous damage. If we are to deter as well as deflect hidden attacks, America needs to detect anomalous behavior on land and on the high seas, as well as in the air as is now being done, and investigate it thoroughly in every instance, being ready to react appropriately as required. This is true in both in foreign waters as well as in the oceanic areas contiguous to America and its protectorates. Only the Navy has the ability and power to "reach out and touch" someone in mid-ocean, before they get within range to deliver a weapon of mass destruction. Thus, the Navy is critical to the entire effort, from far inland overseas to foreign shores, to mid-ocean to our near seas to our ports.

The football analogy works well here. Some believe we can win all our games by "rushing the passer" every time; hopefully sacking him before he has time to deliver "the bomb" (literally in this case). However, not enough thought in the Navy has been given to the problems that arise if the "quarterback" does get loose or a "running back" sneaks around and heads for the end zone at full speed. Clearly, we do need to "rush the passer," but the Navy and the other agencies charged with maritime security better have second and third options ready to go, "linebackers" and "safeties" trained and ready to tackle "the guy" as he heads for what might today be an easy touchdown, if tackling our opponent in his own backfield fails. The public will never forgive the Navy if a major attack occurs in our ports and it appears the Navy could have been instrumental in preventing it, if it had been ready. The fact that the actual responsibility for port security now rests to a major degree on the USCG and the US Customs Service will not absolve the Navy from receiving a large share of the blame from the public. The credibility of all three organizations will be damaged, but the Navy will take the biggest share of the blame and its credibility will have been severely damaged, possibly beyond repair. The forward strike mission, the major jewel in the Navy's crown, could well be transferred to the Air Force in toto as a sign of the lack of that confidence. The Navy needs to work a great deal closer at all levels with the other agencies which have any part of the maritime border security mission to be ready, should the need arise.

One of the major components of the required readiness is the ability to track all ships inbound and outbound from our shores, well before they arrive. We do it for aircraft, traveling many times faster than ships, but not for surface vessels. Because of the significantly increased security measures enacted in the airline industry worldwide many, including this author, believe the next major attack on America will come by sea, whether it is in a container or within the hull of a smaller ship remains to be seen. However, containers are now receiving substantially increased scrutiny; thus, it is entirely possible the next attack will come from a smallish vessel, one that masquerades as a fishing or research vessel, or large pleasure craft or other large working boat. Indeed, in the late 1930s Albert Einstein foresaw the possibility that the first atomic bomb detonated against our country would be delivered by ship.

Little has changed regarding maritime border surveillance since then, but nuclear weapons have gotten substantially smaller and much more lethal. Even more likely to this author is the idea that several hundred pounds of the anthrax or other deadly compound Saddam Hussein keeps telling the world he has destroyed, but cannot provide proof thereof, will be delivered simultaneously to several US ports via relatively small vessels rigged much like Timothy McVeigh's truck. The few bags on anthrax (?) resting on its foredeck are sitting on a steel plate and are rigged to explode once they are thrown several hundred feet in the air. The ships detonate themselves within a few minutes of arriving in the ports, or if interdicted, in the immediate vicinity of their target harbor and its nearby city. The actual damage by the detonation may be minimal but the terror generated by even the rumor that bags of a lethal substance such as anthrax were blown into the air by such an act would spread terror from one coast to the other and virtually bringing the commerce of the US to a halt for months. It is truly a nightmare scenario.

The only way to accomplish the required level of ship tracking which might deflect this type of attack at an attainable cost is by building the maritime equivalent of the North American Air Defense Command (NORAD). The Identification Friend or Foe (IFF) radar transponder is the core technology for the effective operation of NORAD. It allows radar operators to quickly identify the vast majority of air contacts, to sort the many valid, known tracks, from those few that need additional scrutiny. Because ships operate beyond line of sight of shore-based radar, IFF transponders will not work. However, engineering studies show that ship-borne satellite communications transceivers, called satellite communicators (SC)s by some, could provide the same functionality as radar IFF transponders.

This author has written a paper outlining how such a system, called the Maritime Traffic Tracking System (MTTS), could be built from both systems and concept of operations viewpoints. However, the actual interfacing and fusion of the requirements and resources of a multitude of governmental agencies across at least four departments; those of Defense, Homeland Security,

State, and Justice requires a much more detailed investigation prior to full implementation. The core capability will be a maritime traffic tracking system of some type; however, the competing needs and wide variety of resources available must be carefully integrated to minimize the impact on current operations and maximize the impact of the new capabilities and organization on existing and emerging requirements. The competing requirements, coupled with the wide variety of capabilities available, call for thoughtful systems engineering to develop the maximum utility of the new system.

Much of the required infrastructure for a maritime traffic tracking system Maritime Traffic Tracking System is already in place. Indeed, the US Customs Service operates a facility at March Air Force Base, Riverside, California, called the Air & Marine Interdiction Coordination Center (AMICC). It currently tracks all air contacts inbound and outbound from the borders of the United States as well as much of the air traffic over the rest of North America. The AMICC has raw video provided from many of the air surveillance radars observing the borders of North America. The AMICC's systems also allow it to interrogate the IFF of any of these tracks. There are plans to expand this tracking capability to include even the radars in the interior of the US. At any one time, it has information technology tools which allow it to have approximately 5,000 air tracks on its screens. It routinely investigates between 2,600 and 3,000 anomalous tracks a day. Assisting it in this task are information technology (IT) tools that allow them to review the current flight plans, as well as cargo and passenger lists for any flight it is tracking. It also has near instantaneous access to the history of any aircraft it is tracking, including the graphic display of all previous tracks and automatic notification of any unusual activity as well as the names of any passengers of high interest. The AMICC has identified the augmentation upgrades required to allow it to track all self-reporting surface vessels within 1,000 miles of our coasts. Those upgrades are in the $10 million range for facilities expansion and needed software and hardware enhancements, upgrades as well as for additional surveillance consoles. They also believe they could take on the expanded task with those improvements and 15 additional surveillance console operator/analysts.

As is mentioned above, one of the core competencies of the AMICC is its ability to both connect the data contained in current flight plans with actual tracks as well as with historical data on any aircraft such as previous flight tracks and association with high interest persons. This allows for a high degree of anomaly detection when tracking specific aircraft. The USCG mandated advanced notice of arrival (ANOA) required of all commercial ships 96 hours before they enter our ports contains much of the same information as a flight plan. It requires notification of port of arrival (POA), estimated time of arrival (ETA), as well as people and cargo onboard. It also now requires the point of origin and location of the unit when it is making the initial report, 96 hours prior to arrival in port. This last item was included when it was determined many ships were already very close to the port before they were reporting their arrival. The ANOAs are fed into the Ship Arrival Notification System (SANS). The AMICC flight plan analysis software and its associated databases could easily be modified to allow for the ready access of the SANS by surveillance console operators and maritime analysts. Additionally, there are tracking display and decision aids for tracking entities such as Secure Asset Reporting System (SARS) utilized by the ORBCOMM communications satellite system, that already have similar functionality as one of its core capabilities.

The USCG stood up two Maritime Intelligence Fusion Centers. One is at Dam Neck, Virginia. It is charged with providing tactical and operational intelligence to US government units operating in the Atlantic Ocean and Caribbean Sea, as well as the Gulf of Mexico. (and inland waterways in the eastern half of the US plus Texas?) The second one will be at Alameda, California, with tactical and operational intelligence support responsibilities for the Pacific region.

However, the AMICC upgrade and the creation of the MIFC are only part of the solution to the Maritime Domain Awareness (MDA) problem. Additional analytical/operational intelligence support is also required. At a minimum at least one full-time,

24/7, watch person would be required at each of the two MIFCs, Atlantic & Pacific, as well as the Guard and at the Office of Naval Intelligence's (ONI) civilian shipping analysis section at National Maritime Intelligence Center (NMIC). The NMIC is charged with providing strategic and operational intelligence as well as support to the tactical intelligence units, both at home and abroad. The interfaces to the military and law enforcement (LEA) units tasked with supporting these operations and to each of these organization's higher headquarters the several law enforcement agencies directly involved would also need to be defined and developed. The command & control relationships, as well as technologies, would also need to be established. Each organization has a different task and brings different resources to the problem. Each also has a valid claim on the output of a MTTS. Additionally, there are large functional, legal, operational, and technical questions that need to be addressed in the correct political framework before the optimum solution is derived. For instance, what level of increased operational tempo is necessary on the part of USN, USCG and USCS tactical units? What is the utility of MTTS, as well as its impact on tasking, for our wide area sensors such as our national technical systems, the acoustic surveillance system, and the over-the-horizon radar systems? Is the manning level correct for this mission at the Coast Guard's MIFCs and at the NMIC?

The idea of a maritime NORAD has had a good bit of discussion since 9/11 but many believe it would be both too costly and not effective enough. This author believes that building an effective maritime traffic tracking system, and an associated maritime interdiction system, is neither impossible nor too costly, especially when one considers what is at stake. A successful nuclear, chemical, or biological weapons attack on one or more of our ports could well be significantly more devastating to America's economy than the attack on 9/11. A successful coordinated set of attacks on several of the largest West Coast ports, not totally beyond the realm of possibility, could have an almost unimaginable impact on the American economy, possibly crippling it for the foreseeable future. Even a small attack could have political, economic, and social impact all out of proportion of the actual damage inflicted, spreading fear and uncertainty across the entire social fabric not just of America, but the entire world. The government of the United States, especially the United States Navy, would be blamed for the failure, whether or not they were actually culpable.

Defense of our maritime assets, both here and abroad, is the task of the Navy in the public's eye, even if the actual responsibility for a particular segment is assigned to the Coast Guard, the Customs Service, or whomever. Just as the entire Intelligence Community was damned by the Congress for its failures to coordinate prior to 9/11, so too would be all these three services if a successful significant attack occurs in our ports, however, none more so than the Navy. All three of these organizations need to be working together, from top to bottom, as closely as possible, on the protection of our maritime assets. They need to jointly develop the proposed system as well as also be completely interfaced into the other maritime stakeholders such as the Immigration and Naturalization Service (INS), the Drug Enforcement Agency (DEA), the Environmental Protection Agency (EPA), the Maritime Administration (MarAd), the Justice Department and the Department of State. The Intelligence Community must be a full participant as well.

This situation is remarkably like the introduction of radar at the beginning of World War II. Both sides knew the other side was working on radar, but it took trial and error and the loss of many valuable combat units well into the Korean War before the full advantages and liabilities of radar were understood. Those lessons were largely ignored and forgotten at the end of the Korean War and we lost many more aircraft in Vietnam before we fully relearned the lessons of the Korean War regarding radar. Obviously, the real advantages radar gives an air defense system and the necessity of countering those capabilities, especially at the very beginning of an air campaign, are well known. Less well understood are the electronic warfare liabilities of an active air defense. The fact that the United States has lost so few aircraft to air defense systems since the Vietnam War indicates we may have finally learned those lessons. It was about time. Hopefully, we will learn the lessons of 9/11 in far less time.

The WALEX process

It is with these thoughts in mind that it is asserted that the systems engineering process at APL called the Warfare Analysis Laboratory Exercise (WALEX), could be of significant utility in the introduction of MTTS and a maritime NORAD. The WALEX process was designed to facilitate the tailored, structured examinations of the impact of science and engineering advances on military operations, especially air and anti-air operations. Both the process and its facility, the Warfare Analysis Laboratory (WAL), have grown and matured over a number of years. They are the product and outgrowth of 20 plus years of systems engineering and technical analysis. The current facility is the fourth generation of specifically designed facilities to assist in the analysis of problems with a large technical component. As indicated above, the process was originally designed to further the analysis of the impact of science and engineering advances on military operations, but it has since been used to investigate a wide range of activities impacted by advancing technology such as medicine, logistics and traffic management.

Just as no one had a clear understanding of all the ramifications radar (both its use and its exploitation) was going to make on warfare at the onset of World War II, no one person, or even any specific group, has a clear understanding of the impact of weapons of mass destruction in the hands of terrorists, or the capabilities a maritime NORAD might give us to counter that threat or its effect on future maritime surveillance operations. Because these problems are so multifaceted, the understanding of the impact of MTTS on maritime surveillance and national security needs a multi-faceted approach. Once a terrorist attack has been perpetrated it will be too late to start bringing the correct coalition together to try to fully grasp the opportunities provided and solve the problem. Opportunities to counter and engage hostile forces and law breakers could be irrevocably lost, and lost at great cost. Unbelievably valuable assets, including port infrastructure, ships, and, even more valuable, people's lives could be lost before we get it right. We need to start to deploy a MTTS now, and we need to develop the capability to fully explore the ramifications of MTTS as a principal component of a maritime NORAD now, before it is fully fielded. Scientists, technologists, intelligence specialists, law enforcement professionals, military operational experts, and even legal and political specialists need to be gathered to explore the possible impact of MTTS and develop concepts of operations and tactics in a benign environment such as a specially equipped laboratory, prior to the full introduction of MTTS. These exercises and their associated LOEs need to be structured to gain the most from them, or significant wasted time, money and opportunities will result. The structuring of both WALEXs and LOEs is much more difficult than is widely recognized but the structuring of both together can be highly beneficial.

A discussion of the background of the WALEX and why it is the right tool to examine this complex problem is required. For ease of discussion, the term WALEX is going to be used to discuss this process, although it now goes by several other names, including technical seminar war game, technical war game and the like. It is called a Warfare Analysis Laboratory Exercise or WALEX, after the room in which it was developed at APL, the WAL, but it is much more than a room. The same basic WALEX process is now used, in both civilian and military endeavors, to examine multi-component, cross-discipline problems. It is often used as the requirements definition phase of a project, the initial step of the now widely used systems engineering process. This paper will describe in detail how that WALEX process, either at APL or some similarly equipped facility, could be structured to affect a seamless, and hopefully painless, introduction of a maritime NORAD. A series of properly structured WALEXs would be of significant assistance in defining the requirements for, and designing the successful integration of, multiple systems into a maritime NORAD.

The Warfare Analysis Laboratory (WAL) was designed to facilitate the systematic technical dissection of a multi-dimensional problem with a significant scientific component. However, it is the process, the WALEX, and the trained WAL staff, assisted by the laboratory's Joint Warfare Analysis Department (JWAD) analysts and facilitators, which are the key to the successful development of a technically sound solution to an operational problem. In the case of MDA and the need for a maritime NORAD there is also a huge political component to the problem and a WALEX can also be used, at the same time, to generate consensus and understanding across the operational and political boundaries to establish the cross discipline buy-in which is so important in this instance.

Many institutions have built decision support centers to conduct structured analysis meetings. APL's WAL has evolved over the past 20 plus years, from a small meeting room with a single screen to a large four high-definition video screen multimedia meeting room with an electronic seminar system (ESS) with 53 stations and provision for over 100 additional participants. The ESS is for the participants to record their inputs in real time for later analysis. There is also a gallery for additional participants, who can input comments into the system ex post facto. However, a WALEX, correctly done, is much, much more than a technically focused and structured seminar war game in a specially configured conference room with superior presentation, decision assistance, and data capture tools. Indeed, the WAL is a great place to have technical exploratory meetings because of its layout and the technically sophisticated multi-media tools it has available, but using the WAL that way is like using a race-prepared Porsche to go pick up groceries. It can do that job too, however, it is a lot more fun to open it up and really see what it can do in the hands of a skilled "driver."

To carry the racing Porsche analogy a bit further, while it is the driver that actually drives the car, it is the team that actually prepares the car that makes a critical difference in the result of whether the Porsche wins or loses. Both, together, the driver and the team, determine the outcome, whether the car (and the team) wins or loses. Their fates are interlocked. So is it in a WALEX or, for that matter, any structured group analysis. You need a trained team to prepare for the task, to bring focus and experience to the problem at hand, and a trained facilitator (driver?) to keep the group moving forward to the goal at best speed. Many places now have the facilities, but very few have the trained team with skilled facilitators.

A WALEX steps through a series of defined events to bring the problem under investigation into focus, using whatever lens is appropriate for the task at hand at that specific point in the process. Those steps have been tried and refined for a number of years. Under the guidance of an experienced facilitator, the team prepares a WALEX, going through each of the defined steps in order. It decomposes the problem into its political, operational, systems, and technical components for further examination to any desired level of detail, down to and including the definitive technical elements of a problem. Engineers refer to this as "the first principles" of physics. Political scientists have similar core concepts. The WALEX process can provide for the detailed, focused exploration of the engineering and technical challenges to be faced in implementing the various operational and systems solutions under consideration for a specific system or mission area. It can also be used to look at all the other aspects of the problem, including the necessary tactics, techniques, and procedures as well as the required skill and knowledge set of the operators and decision makers directly involved. It can also examine the training required to provide the knowledge and wisdom to carry out the required missions. The political aspects of the problem could also be factored into the discussion for full consideration. Modeling and simulation at several different levels of abstraction are normally employed throughout the process. The MDA/maritime NORAD problem is a significant enough departure from what most operators, be they LEA or military, are accustomed to dealing with that it will demand this level of examination. One develops understanding on the job, but it is often hard, if not impossible, to back up and replay an operational situation to examine how it could have been handled differently. It is easy in a WALEX.

The first step the team undertakes, in coordination with the sponsor, is Objectives Definition. The WALEX process starts with the need to define what the desired end-state is, what the objectives of the overall program are, and what is expected from both the first WALEX as well as any subsequent efforts. Often the investigative process is broken down into what is expected from each of a series of WALEXs. The same is true of LOEs. For the remainder of this paper, it will be noted where there is commonality between the two processes. It is not uncommon for those objectives to change during the course of the series. It is often very instructive to capture them in the beginning and then look back at the original list as the process evolves. The process next calls for the determination of the desired product, the role of the WALEX (or LOE) at hand and the identification of the salient issues to be examined.

When the objectives are established, and the overall design is in place exercise preparation gets underway. Technical and operational information is compiled, for all involved entities, both friend and foe alike. It is important to understand the most likely threat environment as well as your own capabilities. It is also worthwhile to understand possible excursions from the most likely scenario. Participants with the requisite experience, skill set, or technical knowledge are selected and invited. This step is especially important and is undertaken with full coordination with the event's sponsor. At the same time modeling and simulation support needs to be identified and prepared. It is also in this step that facility requirements are defined. How big a room is required is generally dictated by the number of participants. Often the classification level of the game will require a sensitive compartmented intelligence facility (SCIF) for part of the time to establish an intelligence baseline. The timing of meetings and briefings in the SCIF, which will generally be attended by only a subset of the participants, need to be coordinated so as to not adversely impact the overall process. The tactical displays need to be developed with understandable iconology. With the advent of several commercially available superior graphics packages, this is not as difficult a task as it once was, but attention must be paid to this because not all participants will be familiar with military display systems and a good bit of time can be wasted bringing civilian personnel up to speed on what is being depicted on the screens with military data link symbology.

Once all the above is accomplished it is necessary to conduct a full-up dry-run, be it a WALEX or LOE, to ensure there are not any loose ends and to be sure the exercise gets at the issues under consideration in the most beneficial manner possible. At the end of the exercise pre-play the exercise book is updated to its final form, published, and distributed as a "read ahead." The WALEX (or LOE) is ready to start.

At the commencement of the WALEX the objectives and technical approach are detailed, and the scenario briefed to the participants. These will have been basically defined in the exercise book, but a restatement at the beginning is necessary to gather any last-minute input from the participants. As the scenario is stepped through the key interactions are examined in detail. This is one of the reasons pre-play is important, to anticipate the questions that will arise and develop answers and the necessary accompanying graphics to explain them. Decision analysis of key issues can be provided at this time. Critical issues for further examination will naturally surface during this process. They will be recorded and used to fashion the next iteration of the WALEX series or taken back by the participants to their parent organization as input on where to focus future research efforts. The entire event will be captured by several means. Trained note-takers will record major points and the facilitators will generate a quick look report, often just for their own use. Additionally, the WAL is equipped with a state-of-the-art electronic data capture and decision aid system called the Electronic Seminar System (ESS). It consists of a series of laptops at each of the principal positions around the large conference tables arranged as a set of Vs, as well as at some of the seats on the side. Any individual can make comments in the ESS either anonymously or with full attribution. Generally, there is an expanded agenda embedded in the system and participants will flag their inputs to specific agenda items or as an overall comment. Comments can be displayed for all on one or more of the four large screen displays at the front of the room, if so desired by the facilitator.

As is obvious from the foregoing, most of the work that goes into a successful WALEX (or LOE) takes place before the participants even arrive. This is not to denigrate the absolutely crucial input the participants make to the successful effort during the actual WALEX, but rather to point out the need to gather all the pertinent facts and have them organized and developed as fully as possible before the exercise begins to allow all to have a common, technically and operationally correct, as well as politically feasible, point of departure from which to commence deliberations. This is a major point and one that is not generally recognized by most people contemplating putting on a technically focused seminar or war game and is one of the major discriminations of a well-run WALEX. The scientific component of the problem under consideration needs to be made understandable to the non-scientist participants. The operational framework needs to be made understandable to the scientists. The political context needs to be understood by all participants. How well these processes are done can determine the success of the WALEX. If they are not done well it is exceedingly difficult to have a successful WALEX (or LOE). It is not as easy as it seems. It requires a great deal of interface and interchange between a knowledgeable sponsor's representative (or representatives) with the exercise designers. Indeed, it does take a good deal of homework and requires exceptional attention to detail.

The most valuable way to use WALEXs is to do a series of them in an iterative fashion. A WALEX, properly prepared and facilitated, will break a problem down into "what is known, what is not known and what is believed, but not proven." Possibly even more importantly, it can also shrink, but obviously never eliminate, "what is not known is not known." By knowing these facts future research and analysis efforts can be refocused to fill in the voids. Once the refocused research has filled in some of the voids a new WALEX (or LOE) can be conducted with the new findings of that research inserted into the next evolution. In this way a problem can be moved more quickly to its logical solution.

After each WALEX the exercise results are analyzed, and issues identified and resolved if possible. Once that process is complete the results are published. If the WALEX (or LOE) is part of a series the results may be only an interim step, but at the completion of the series, a final report is published and distributed.

At each iteration of the WALEX it is best to use some of the same experts but also invite other experts with relevant expertise to get fresh input. Newcomers need to be limited to people that can be brought up to speed quickly and can contribute immediately. Otherwise, a great deal of time is spent re-plowing old ground. Read-aheads are the means most often employed to shorten the process of bringing new members of the team up to speed. This is true of even the first WALEX in the series. Read-aheads and being sure that only people who can, and will, contribute are invited to participate are two items that are overlooked only to everyone's detriment.

A series of WALEXs is best interspersed with research, limited objective experiments (LOE)s and then, finally, inclusion in major exercises/integrated fleet experiments such as the Fleet Battle Experiment (FBE) series. Each step should be employed by the involved entities to develop a mutually shared understanding of the impact of a maritime NORAD. That level of understanding would take a much, much larger number of technical conferences to achieve.

The five communities of policy, science, intelligence, law enforcement and military need to be brought together in a coherent, structured manner to examine the potential cross-organizational impact of MTTS and a maritime NORAD on MHLS/MHLD and MDA and begin to prepare for its introduction as both a system as well as an operational concept. These discussions might be initially held at a low classification level, possibly even unclassified but they would quickly separate into sub-groups that would allow meetings at different classification levels.

MDA and the proposed MTTS and maritime NORAD system is a classic operational/technical/political problem for investigation via a series of WALEXs. This is especially true because there are, as mentioned above, many interrelated political, technical, system, and operational elements to the MDA/MTTS/Maritime NORAD problem. Each segment of maritime surveillance operations has both competing and complementing technologies and methodologies. MTTS may well affect many, if not all, of them. The primary function of many of these technologies and methodologies is to acquire and move information from a sensor to an action taker in as timely and complete a manner as is necessary and possible. The data/information needs to pass through a decision maker who must be able to reach the correct decision as quickly as possible. The decision maker needs to have a large amount of data available to him/her in a timely manner and in an easily understood format. That data needs to cover not just the location of a potential target but also provide a correct description of the status of all available forces, both military and LEA, and the political situation as reflected in the rules of engagement in force at that instant in time.

As an example, a WALEX could be structured to step through the Maritime Interdiction Operation (MIO) process as it is affected by MTTS, increment by increment; examining way needs to be included in the decision mix, the system trade-offs/options available at each step of the way. The process can be stopped and re-run to examine the impact of each decision on both the next immediate procedure as well as the overall MDA timeline.

MIO is significantly more complex than is generally believed. Each segment of a MIO operation has both competing and complementing technologies and methodologies. As was alluded to above, the primary function of all these technologies and methodologies is to move information "from sensor to shooter" (boarding party?) in as timely and complete a manner as is necessary and possible. A WALEX (or LOE) could be structured to allow an informed group to step through the various aspects of MTTS and its potential impact on the MIO process increment by increment, examining the tactics, science, and system trade-offs/options available at each step of the way. The process can be stopped and re-run to examine the impact of each decision on both the next immediate procedure as well as the overall impact of MTTS and a maritime NORAD. Stopping and re-running is often hard during a live LOE, and obviously, it is impossible in real world operations.

A series of WALEXs is not particularly cheap to put on, but they are much, much cheaper than at-sea or on-range tests. They are also many times cheaper than losing even one ship or plane involved in MIO, not to mention its crew. It almost goes without saying that a major incident in even one port would cost far more. WALEXs can make LOEs and at-sea tests much more productive and cost effective. In fact, they can pay for themselves many times over just by making real world tests much more meaningful. They could also be used to build a data collection and analysis plan, both for set live tests and as collection guidance to intelligence systems.

It is time to move the implementation of an integrated MDA system, the MTTS and a maritime NORAD to the next step and fully involve all affected elements of our government in developing an understanding on what problems and opportunities are created by a maritime NORAD.

Recommended the appropriate people within the intelligence, political, law enforcement, and military communities get with the professionals at APL, or some other similar institution, who design and lead WALEXs to explore the idea of using WALEXs to define the maritime NORAD problem and develop a technically, politically and operationally feasible road map for its implementation. It is past time to bring all parties into the discussion and devise a solution for the defense of our porous maritime borders. It would be a win-win-win proposition for all Americans. We owe the American public nothing less.

Appendix Three
Meeting Maritime Security Challenges
in the 21st Century

*Using Capabilities Based Assessment to Develop and
Implement the National Maritime Domain Awareness
System Implementation Plan*

The National Strategy for Maritime Security, and its supporting National Plan to Achieve Maritime Domain Awareness, written to fulfill the requirement of the joint National and Homeland Security Presidential Directive, NSPD-41/HSPD-13, "Maritime Security Policy" reflects the maritime security challenges of the 21st Century. It directs the development of a sustained, continuous collaborative effort across the entire government of the United States, working with state and local governments, private organizations and foreign partners to provide a complete system to develop critical intelligence and information from all sources, classified and open source. It goes to maritime operational commanders at all levels in a clear and concise manner in time to allow them to make the correct operational decisions and initiate the correct tactical action.

It is clearly recognized by all participants in this development effort that success for the MDA Implementation Plan, and its attendant Investment Strategy, execution demands unprecedented cooperation and information sharing among governmental agencies and organizations at all levels as well as the maritime industry, and international partners. The system requires an enhanced collaborative information environment (CIE) made up of information from human intelligence collection, defense, law enforcement and private organizations, and the integration of existing and emerging sensor technologies, analyzed and fused in an operator user-definable common operating picture (UDOP) operating in a multi-level security environment. Users with the highest clearance level would have access to all information, with those at lower levels of security clearances only having access to information appropriate for their level of clearance. Provisions for special access programs would also need to be accommodated. The technology exists to build such a multi-level system; however, getting all parties to agree to build it and develop a real CIE within the total community of interest (COI) is the real challenge. Getting all to agree to change their policies, procedures and, in some cases, the governing laws, is the real problem. Indeed, we do need to improve many aspects of our technology, but that is the easy part; getting buy-in from all of the participants to such an extent that they are willing to return to their parent organizations and move to get policies, procedures and laws changed is the hard part.

This multi-dimensional task requires an equally sophisticated implementation effort to ensure the requirements of all stakeholders are fully considered and maximum "buy-in" is achieved across all departments, agencies, and all other entities within and without the government. Thus, even the development of the implementation plan itself calls for the use of proven systems engineering methodologies and a structured analytical approach to the multi-departmental, complex problem of building an effective national Maritime Domain Awareness system.

The Operations Research paradigm detailed below is useful, but the effort undertaken here is much more complex due to its multi-departmental impact, if nothing else.

With those thoughts in mind, it is proposed that the implementation plan itself be developed by using a series of structured technical, systems of systems level focused seminar wargames commonly called WALEXs in defense analysis circles. WALEXs have been successfully used a number of times to develop overall and breakout plans to build a wide variety of complex systems, including the National Ballistic Defense Plan, the National Space Security Communications Architecture, the Community Emergency Preparedness and Response Plan (CEPAR) and many other complex systems. They are designed to bring the results of detailed, cross community and cross discipline systems engineering efforts to groups of analysts and/or decision makers to allow informed determination of the way ahead, and very importantly, foster informed cross-community understanding of the realities facing each participant's community, be it technical or political.

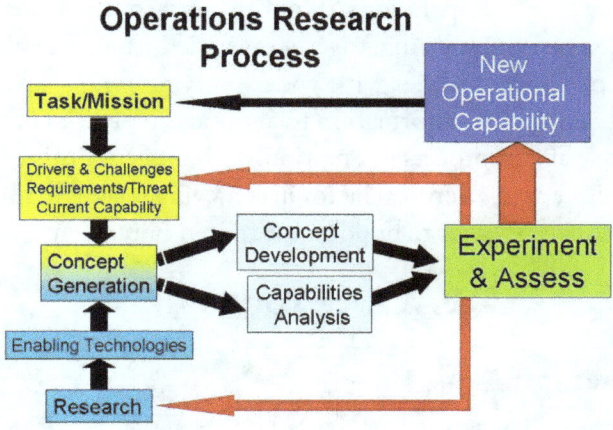

Figure 1: Illustration showing conceptual planning and recommended action to achieve Maritime Domain Awareness. Courtesy of author

The term WALEX refers both to a refined system engineering process and to a specific tool in the form of a specially configured large room, the Warfare Analysis Laboratory (WAL), developed over 20 plus years and several iterations at Johns Hopkins University's Applied Physics Laboratory (JHU/APL). Both the process and the WAL have undergone substantial development over the past quarter century to the point that a specifically structured exercise (Ex) in the WAL has come to be called is a WALEX, but not all meetings in the WAL are WALEXs and certainly not all WALEXs happen at APL's WAL. The MDA Summit used the WAL and its trained staff but did have have the rigor of a true WALEX. It did use many of the same techniques in preparation.

Figure 2 outlines the steps in the WALEX process. This process has been used to develop overall enterprise architectures and their associated subsystem structural design, and brought focus to the development of pertinent data and interoperability standards by doing trade-off studies supported by focused modeling and simulation. This process can work on sensor and data fusion/ mining/correlation, and data display and decision aid (COP) technology, as well as procedure development for such things as a Concept of Operations (ConOp) and its associated parts. WALEXs routinely bring together parities from across a wide variety of responsibilities with widely varying points of view for mutual exchange of information to allow acceptance and buy-in to the development process in question. This aspect of this process is near-unique and is a critical factor in achieving success in the multi-dimensional endeavor to build an integrated Implementation Plan.

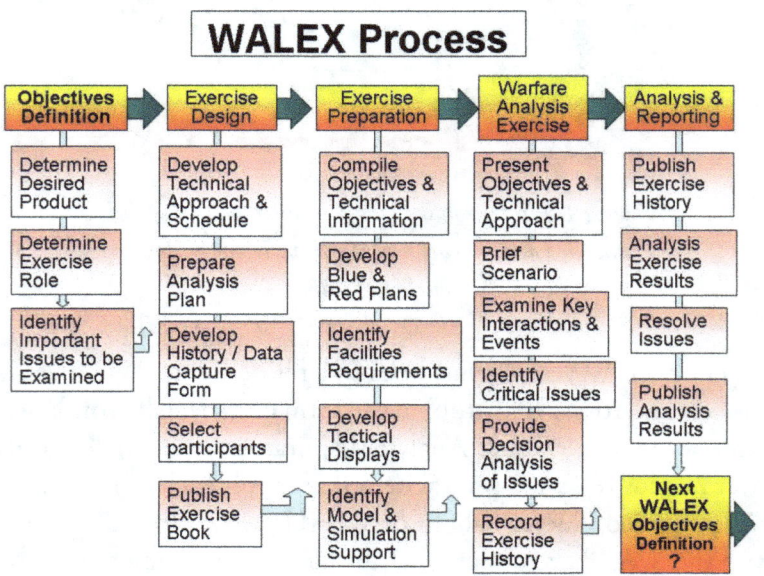

Figure 2: WALEX step by step. Courtesy of author

Once a basic plan is built various aspects of it can be tested and exercised in carefully crafted limited objective experiments (LOEs) such as those being developed by the Naval Postgraduate School and the Navy's NetWarCom's Trident Warrior experimentation series. However, much work needs to be done before we reach that level. This dual track of systems engineering driven WALEXs and LOEs would allow the maritime surveillance community to "build a little, test a little" as it develops the complete system. This deliberate approach would allow for the most efficient planning for, and implementation of, the seamless integration of the multitude of requirements and assets involved, while quickly giving us a vitally needed maritime border surveillance system. It would also go a long way to developing cross community agreement and understanding of the role of each participant.

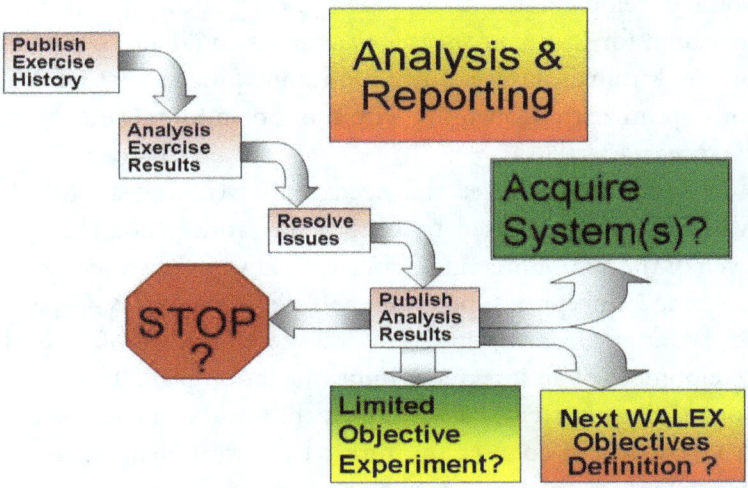

Figure 3: Illustration of WALEX analysis and reporting. Courtesy of author

A sophisticated process such as this is necessary to gain a full understanding of the utility (and limitations) of greatly expanded inter-departmental Maritime Domain Awareness and border surveillance systems and their impact over the complete operational spectrum both within each affected department as well as the interactions of specific actions and interactions across departmental boundaries. This process can also be adapted to fulfill each element of the Department of Defense Architecture Framework (DODAF) and the Joint Capabilities Integration and Development System (JCIDS) process.

The actual interfacing and fusion of the requirements and resources of a multitude of governmental agencies across at least six departments; those of Defense, Homeland Security, State, Transportation and Commerce, and Justice is daunting. The multiplicity of departments, each with its own needs, requires a detailed investigation prior to full implementation. The WALEX process could be used, either at JHU/APL or some similar facility with a trained staff, including, most importantly, trained facilitators, to develop the investigation in a structured manner. It would examine Maritime Domain Awareness system (MDAS) technology and interrelated options for concepst of operation. It would develop a much fuller understanding of the opportunities generated for MDA by specific systems. It could also be used to assist in formulating tactics, techniques and procedures (TTPs) to both optimize this capability within our own forces and counter the threat.

In summary, to meet the needs of developing an MDAS, WALEXs could be used to further the work generated by all seven of the Senior Steering Groups working groups and the now on-going Concept of Operations and Requirements & Capabilities Working Groups, with its several sub-working groups, in the development of an integrated implementation plan. The process would bring a better focus on the capabilities, both technical and legal, needed to counter the threat to critical elements of our maritime infrastructure posed by potential opponents; as well as to assist in the optimization of the capabilities of our own resources in a counter-terrorism role and other governmental maritime roles such as counter-narcotics operations, environmental protection, off-shore fisheries surveillance, illegal immigrant smuggling prevention, and safety of life at sea operations such as search and rescue. New technologies and policies could be introduced and examined in both an operational and technically correct scenario. Legal and political dimensions, both foreign and domestic, could be examined in detail as well. LOEs could be used as needed to test the results in a real world environment and to train the various elements of the detection-analysis-decision-action cycle.

The WALEX process proposed as the principal tool for the systems engineering examination of this problem was initially designed to allow for the structured examination of complex problems with significant technical aspects, specifically, the myriad challenges of naval air defense. However, the WALEX process has expanded over the past quarter century and is now routinely used to gain a much fuller understanding of the requirements and impacts across all aspects of doctrine, organization, training, material, leadership, personnel, and facilities (DOTMLPF). WALEXs are now routinely used to gain insights in many warfare areas, as well as numerous other less highly technical scenarios such as disaster relief, updating highways, and biomedical research. Since 9/11 there has been Community Emergency Preparedness and Response (CEPAR), a major effort in the National Capital Region to develop disaster preparedness plans.

It should also be noted that, while JHU/APL developed the WALEX process, other organizations have also developed expertise in staging these complicated efforts and anyone contemplating staging such an effort should look at all alternatives.

Appendix Four
International Collaboration Is THE Silver Bullet

(C-SIGMA) Collaboration in Space for International Global Maritime Awareness

Following is a memorandum that I distributed widely in the maritime security community in 2006 and later updated.

Prologue

There is no silver bullet, not now, nor in the foreseeable future, but all maritime nations of the world, working together, can make the seas much safer and more secure from wrong-doers, be they smugglers, polluters or pirates. One of the primary steps the nations could take would be to create a global space partnership (GSP) initially focused on the maritime domain. Such a concept has been under informal discussion for some time by many people. I have named the maritime focused portion of the GSP, Collaboration in Space for International Global Maritime Awareness (C-SIGMA). It is certainly recognized that a GSP would have a much broader capabilities than just the maritime domain, but many have recognized the critical vulnerabilities of our maritime assets and the potential huge economic impact their loss could generate, and thus the pressing need for much better awareness of the maritime domain. Indeed there is a parallel effort already under way in regards to protecting the global environment.

The Group on Earth Observations (or GEO) is coordinating international efforts to build a Global Earth Observation System of Systems (GEOSS). This emerging public infrastructure is interconnecting a diverse and growing array of instruments and systems for monitoring and forecasting changes in the global environment. This system of systems supports policymakers, resource managers, science researchers and many other experts and decision-makers.

But they obviously are not focused on global maritime awareness, and they have not been particularly interested in

186

joining the C-SIGMA effort, although the author has approached them several times.

Increased awareness starts with better ocean and coastal surveillance. This point has been made twice recently. First, the recent wide-spread recognition of the fact that piracy is alive and well in the 21st century and is a growing, not diminishing, threat. Additionally, the terrible attack from the sea on Mumbai, India in 2008 by just a few men paralyzed that multi-million-person city and has also brought a lot of attention to the need for better maritime awareness. The author has studied the situation in depth and has come to believe unclassified space systems will play a major role in any effective maritime awareness system. It is realized that space systems cannot do it all and collaboration and coordination with terrestrial systems as well as the mining and analysis of semantic data contained in hundreds, if not thousands, of databases is also needed. Likewise, it is recognized that coordination is needed down to the responding tactical units, but those are wholly different issues and will not be addressed here, except to be acknowledged.

Oil may be the world's lifeblood, but oceanic commerce is its backbone, if not the entire skeleton. The horrendous attack on the United States on 9/11/2001 served as a wake-up call in many venues, not just in the United States. The maritime entities of the world, military, civil, and private alike, looked at their situation in the new reality and quickly understood their vulnerabilities and the potential consequences. Since that terrible day a number of national and international organizations have addressed how to protect their maritime assets, both individually and in growing numbers, collectively. Most saw increased Maritime Domain Awareness (MDA) as of first importance to the smooth functioning of commerce on the world's oceans, the crucial supporting frame of the world's economy, and crucial to their national interests. The growing realization that piracy is alive and well in several areas of the world, and the devastating sea-borne attack on the civilian heart of Mumbai have reiterated the need for maritime vigilance. There is also a growing realization that misuse of the oceans can lead to significant environmental damage and huge loss of natural resources. Thus there is little wonder a wide variety of organizations either have or are developing systems and concepts of operations (ConOps) dealing with regional, if not global, maritime awareness.

The potential unique contributions of current and planned space systems, owned by a wide range of nations and available to many others, to international global maritime awareness is a subject of growing interest to many. However, in order to understand the true uniqueness of those contributions the background needs to be set first. Many individuals and organizations that have closely studied the problem realize that no one country or even any existing collection of countries has the stature, breadth and depth to successfully organize a meaningful coalition to protect oceanic commerce, the maritime environment, and the broad range of individuals that use the maritime domain for a multitude of endeavors including profit, conveyance, and recreation. They realize it will take international collaboration and cooperation on an unparalleled scale to provide this protection and assure the safe and secure use of the world's oceans. The only organization that has addressed a task similar in scope is the International Civil Aviation Organization (ICAO) and that effort took almost 40 years to reach full functionality after the need was first articulated. Many believe that because the maritime domain has been an integral part of the world's commerce and conveyance systems for thousands of years it will be much harder to create the needed organization to regulate it. This paper addresses some of the technological aspects of the needed collaboration, but we all understand the political aspects of forging that collaboration are the real challenge. Still, a tangible goal, with both technical and policy aspects to work toward, will assist in focusing the political and policy discussions. Building C-SIGMA, the collaboration of the international community to build a universal maritime awareness system for the world using space systems as its backbone, is one such attainable goal. The two critical segments of that universal awareness are shared surveillance assets and a universal common operational picture (COP). We will focus on the why and how of developing the required surveillance assets and let others address the COP.

Indeed many are already working on the COP issue with such programs as Cooperative Nations Information Exchange System (CNIES), Virtual Regional Maritime Tracking Center-Automated (VRMTC-A), Regional Maritime Awareness Capability (RMAC) and the Maritime Safety and Security Information System (MSSIS). Many fewer are looking at the front end of the chain, the sensor end. One needs to look at both, in balance. We will try to restore that balance here.

Most of the MDA ConOps referenced above assumes some form of layered zones of surveillance and defense, from well offshore, to point defense of high-value targets within ports and adjacent waterways. Generally there are one or more zones between those two zones including approaches and coastal zones. The end game, the protection of the high value targets, is a major goal. Those high-value targets include not just significant ships, but also port infrastructure or other targets of high economic, political, or military value. These include power plants, sewage treatment facilities, chemical plants, critical bridges, historic monuments, and the like. This greatly compounds the problem, but every bit of defense helps. Early warning is critical, although the heretofore emphasis on port surveillance systems indicates many do not understand the criticality of early warning, and the need for it to begin far offshore, if not actually with the surveillance of the supporting shore infrastructure across the seas. The Columbian drug smugglers certainly know this. They go to extraordinary lengths to hide their preparations to ship drugs to the US in a variety of means because they know we can track a suspicious ship for great lengths, if we can identify it is of special interest at its point of departure.

Many different groups have studied what terrestrial collection systems (platforms and sensors) are needed to support the core MDA ConOps and what technology is available or will be in the near future. Thus whatever specific MDA ConOps plan is finally agreed to by all concerned, the basic technology to carry it out is reasonably well understood. The numbers of one collection system or another, and where and how data will be fused and analyzed, may change slightly. And so may the display and the decision aids and the decision-making sequence. But the basic technological choices remain pretty much the same. C-SIGMA is a paradigm shift in that until very recently few have considered space systems in this manner. Indeed, in the realm of international collaboration the output from such surveillance systems could be put to very different uses by each of the international participants.

Each of the studies referenced above have basically concluded that there is no silver bullet. No one system can do it all, even in a single zone, much less across all zones of defense. Maritime Domain Awareness requirements span areas from coastal and harbor surveillance and warning to persistent and pervasive surveillance of the broad ocean area. The bottom line is that we will need "systems of systems" in each zone. There is no doubt that much can be gained by netting what we now have to build a collaborative information environment, with a user-definable interface, to arrive at a robust user-defined operational picture, tunable by each user to their particular needs. As indicated above, much has already been done in that area over the last few years. The hurdles to be overcome in this area are now much more policy derived ones rather than technology challenges, but the fact, largely ignored by many, is that if we are to provide persistent and pervasive surveillance of all the areas needed to establish Maritime Domain Awareness, we will need both better surveillance systems and more of them. Doubtless, we also need the means to process, fuse, analyze, display and disseminate all available data; make accurate decisions; and interdict any suspicious vessel before it enters any of our ports or approaches anything of value to us or to our allies and partners, but we need more information at the front end of the Detect-Analyze-Decide-Act (DADA) chain if we are to be successful in most scenarios.

Let us momentarily go to the end game of the scenarios and acknowledge how important it is. Indeed, to build a warning system that can Detect and Analyze, without connecting it through to a robust interdiction capability just means that someday, somewhere, someone is going to die "all tensed up, rather than just surprised" to quote RADM Chuck McGrail, an US Navy fighter pilot and old friend of the author's, now deceased. However, we need to focus on the technology needed to detect, identify and track (D-ID-T) vessels well offshore because many of us working in the trenches believe that is where the greatest needs are. That is also where there are the greatest opportunities for international collaboration, and that is a primary focus of this paper.

The types of sensors currently within ports and in coastal areas such as radars, various types of cameras, and self-reporting systems including the Automatic Identification System (AIS), and other transponder-based systems, are well known. Acoustic sensors and other non-traditional sensors, such as the passive coherent location sensor (PCL), which exploits the reflections of the emissions of non-radar transmitters, such as TV and radio, to determine an object's location also have roles here. However, this paper is primarily focused on just the technology needed to detect, identify and track vessels well offshore. We believe that can best be accomplished by space systems, many of which already exist in the commercial realm.

Far-reaching technology

Several countries now have highly classified ocean surveillance means, but no one has suggested they are adequate to the task of providing persistent and pervasive oceanic surveillance of even a single ocean. Indeed, that point has been made clear in recent months by the lack of effectiveness of the anti-piracy efforts. Classified systems are certainly part of the whole MDA system, but only part, for a host of reasons, not the least of which is their limited numbers due to cost and their limited availability due to higher priority tasking. One should not over, or under, sell what the classified national systems bring to the table, but there is a demonstrated need for better open source ocean surveillance.

It is relatively easy to identify and track vessels that want to be tracked and are complying with international reporting require-ments such as Long Range Identification and Tracking (LRIT) and Automated Mutual-Assistance Vessel Rescue System. However, in the broad ocean area, there is a need for surveillance of vessels to detect bad actors. In fact, some of these bad actors are not emitting at all. This is also true of the myriad numbers of smaller vessels, now openly acknowledged to be a primary threat. Some of these ships are not emitting because of equipment malfunctions, some are just careless, but some are engaged in illicit activity such as people or contraband smuggling, including illegal drugs and arms. Others are engaged in resource theft, such as fish poaching. Indeed, each year billions and billions of dollars are lost to states that have their legal fish grounds plundered. Another type of theft is that of oil and gas. Rogue ships run by thieves steal crude oil from well-heads at off-shore platforms. Crooked officials look the other way due, in part, to the fact that it is currently hard to detect these sorts of illegal activities. The requirement to track vessels as they approach our coasts is generally acknowledged and a number of technologies and changes to methods of operation have been proposed to accomplish D-ID-T on these vessels. These changes are nearly all upgrades to existing systems and methods. There is one technical exception, a special type of Passive Coherent Location (PCL), but we will get to that. Using a consortium of commercial space systems to continually monitor specific areas of the world's oceans on a regular basis would be a major paradigm shift.

Furthermore, this would allow governments to accomplish a perceived need to change their mode of operation from being reactive to being proactive. This means that a sensor must be focused on the area of interest (AOI) for a large percentage of the time, if not continuously, in order to have ready situational awareness of an area of interest (AOI) regardless if there are targets or not. Basically, one does not know whether anything of interest is happening unless one is looking. Developing baseline time histories of what is normal operations in AOIs is critical to understanding what is normal and what should be considered an anomaly and perhaps suspect.

Let's look at the surveillance capabilities and supporting systems now available on the open market. First we want to acknowledge there are other systems besides the aforementioned space systems that can assist in persistent ocean surveillance. One of those tools in this regard is over-the-horizon radar. Currently the United States owns three active Relocatable Over-

The-Horizon Radars (ROTHR) being used primarily to provide air surveillance of the southern approaches to the United States. Using sky wave bounce techniques, ROTHR has a range of some 2,500 miles. ROTHR has also demonstrated a capability to detect surface craft but has a negligible R&D budget to further develop this much-needed capability. The Australians have a similar system, looking north and they have an extensive R&D effort underway to make this system capable of surface surveillance. There is an ongoing joint US-Australian project arrangement studying how a better over-the-horizon radar system could be developed. Currently there is a proposal to conduct a Joint Capabilities Technology Demonstration (JCTD) on the ROTHR to examine and validate new technologies for emerging threats. These efforts show substantial promise.

Another system for broad ocean surveillance is large sonar arrays. Long-range sonar detection of surface traffic has long been understood, but our current system, the Integrated Undersea Surveillance System (IUSS), is oriented in such a way that it will not provide complete optimal coverage of the areas of interest and the cost to modify/update/reorient it to provide such coverage is a budget-buster. Advanced sonar systems deployed as tripwires in certain high-interest areas such as in the Florida Strait; in the Mona Passage between the Dominican Republic and Puerto Rico; off Brownsville, Texas and San Diego, California may have high utility as part of a system of systems, but solving the radar surveillance problem must have first priority. There are many places around the world where a combination of sonar with an OTH radar system would be very useful. Their usefulness could be expanded significantly if their data were fused with that from commercially available space systems.

I believe the most promising class of systems for pervasive ocean surveillance is that provided by satellites operated by a broad range of US and foreign governments and commercial vendors. As we have indicated earlier, no one system or even type of systems can do it all. This is true even when considering the most sophisticated space systems. No one system does it all. Indeed, there are at least four basic types of space-based systems that need to be used in conjunction with each other in this process; six, if you count weather and navigation spacecraft. Two of the four employ active sensors: synthetic aperture radar satellites (SARSats); and Electro Optical/Infrared (EO/IR) Imaging satellites

The other two are based on communications systems: Individual Transponders linked to Communications Satellite (e. g. Iridium, ORBCOMM, etc.); and Automatic Identification Systems (AIS)- a system originally designed for collision avoidance and safety of navigation but more and more coming to be used as a primary ship tracking system.

Let's look at these four types of satellite systems. An example of the first type of system, the Canadian government currently operates two SARSats in a public-private partnership with McDonald, Dettwiler & Associates. It launched the first one in 1996 and it has been sufficiently successful that a much more capable system, RADARSAT 2, was launched in late 2007. Canada is expected to launch an additional three to six radar equipped satellites within the next decade. These systems operate in five basic modes and at low resolution have very wide sensor swaths. Most, if not all, of the coming three to six SARSats will be equipped with AIS receivers. More on that soon. Germany, Italy and Israel have all launched radar satellites and several other countries are moving that way. Each of these satellites carries synthetic aperture radar sensors that can see through cloud cover and detect vessels and their wakes day or night. They have also developed the software to exploit the products of a range of types of satellites, both theirs and others.

While there are only about 10 current truly civilian space-imaging systems in orbit today, several companies/countries have plans to add more.

The next types of systems, the EO/IR satellites, are operated by a number of countries and companies. Their capabilities have expanded to the point where even a layman can, in many instances, look at one of their images and immediately recognize a specific building, and even, in many cases, tell which types of cars (trucks, sedans, convertibles, etc.) are parked in its lots.

The third type of satellite we need to consider are the communications transponder systems such as what is carried on the

InMarSat, Iridium, Global Star, ORBCOMM and other communications satellites. Owners of fishing vessels and many other types of highly mobile platforms employ these systems to maintain an active track of their assets for a variety of reasons. As an example, fishing vessels must be able to prove they did not go into restricted waters. Tugboat companies need to be able to track their assets on a near hourly basis for business purposes. Many other companies employ these self-reporting systems for many other valid business, security, and safety reasons.

The fourth type is a new type of space-based system. The International Maritime Organization (IMO) and the International Agency of Navigation Aids and Lighthouse Authorities (IALA) designed a system to allow vessels to automatically know the names, course, speed, size, draft, and a number of other information items of all commercial vessels within line of sight of their vessel. The author was one of the first, if not the first, to recognize that if that signal could be collected from space it would revolutionize how ships were tracked world-wide. After a number of false starts the author was able to get the United States Coast Guard to fund the development of a very capable Automatic Identification Signal (AIS) collector onboard an ORBCOMM communications satellite. On 19 June 2008 the company launched six satellites equipped for the mission and has announced plans to launch 19 more. Two additional companies, COMDEV and SPACE Quest, have also launched AIS collectors. All three companies are planning to launch more. Thus there is a growing capability, which will only increase, to track the ship safety and identification beacon on a world wide scale. The initial results of these efforts are very promising. As of this date there are eight AIS equipped commercial satellites on orbit and many more are being planned.

Ground segment

What to do with all of this data? Canada, for one, has developed its own ship-detection software called Ocean Suite and the various satellite processors have been designed to complement each other to optimize ship-detection performance. This is just one example: The European community is working to leverage their resources. The two latest COSMOS SKY MED SAR satellites were to be part of the future MUSIS (Multinational Space-based Imaging System), which will combine the resources and space assets of Italy, Belgium, Greece, Germany, France and Spain, while other European countries could still join. The rest of the space-faring world has taken note and discussions are underway in a number of locales, but little concrete has been publicly released to date.

A large player in the area of US civilian space for Maritime Awareness is the Center for Southeastern Tropical Advanced Remote Sensing (CSTARS), at the University of Miami. It, in cooperation with Vexcel Corp. of Boulder, Colorado, has developed Ocean View, a trademarked software program that allows for the rapid analysis of any commercial imaging system to determine if there were vessels imaged. It can generally tell the size, type, course, and speed of the vessel imaged from civilian space borne microwave radar and electro-optical mono, multi, and hyper-spectral systems.

CSTARS, and other organizations, are taking steps to improve processing of the images. It also hopes to gain additional access points by establishing mobile downlink sites in such places as the Azores and in the western United States. This is important because the timeliness of the reporting is dictated by the time between the collection of the data and its downlink to an earth station for processing and reporting. Downlink sites in such places as the Indian Ocean, South America, in east Asia, and other locations across the world to allow for wider collection opportunities and more timely reporting which are a requirement for a truly worldwide system. However, a worldwide consortium of nations would make that a much easier problem to solve.

Maritime Patrol Aircraft (MPA) are another part of the mix. Many countries operate MPAs. Indeed some countries have more than one organization operating them. E.g., in the US the Navy and Coast Guard both have MPAs and the Department of Homeland Security Customs and Border Protection (CBP) operates a fleet of highly modified P-3 fixed-wing aircraft with superb ocean surveillance capabilities. All three organizations, Navy, Coast Guard and CBP, are in the process of installing AIS collection capability into these aircraft. The tactics, techniques, and procedures to make the most of this new capability are only just now being investigated. This could well provide a paradigm shift in the way other US aircraft are outfitted for maritime surveillance. It is recognized in the MPA world that these lessons have global implications, and one suspects those implications are being studied in many places.

The above systems are the primary offshore collection systems, besides ships themselves, in use today. None are optimized for the Maritime Domain Awareness mission, but work is underway to understand how best to do just that, to optimize them to provide much more robust ocean surveillance. Many other technologies are being considered, from medium and high-altitude unmanned air vehicles such as Predator B/Mariner and Global Hawk, to manned and unmanned airships, and combinations of systems. One effort that appears to have great promise is the near real-time integration of the ROTHR with the output from CSTARS, and AIS data collected by the Coast Guard, US Navy, and CBP aircraft and vessels. C-SIGMA is a concept which could be a significant part of this.

One other technology has recently been tested in several venues. That technology is the collection and reporting of ships' AIS and Radar Contacts (SARC). Examples include a system called Neptune built by Lockheed Martin, one called MSSIS + built by the Department of Transportation's Volpe Center, and a third box built by the Naval Research Lab, called Maritime AIS Relay System (MARS). Each of the systems can collect and record the installed AIS and ship's radar for a set period of time and then transmit the tracks and contacts to a shore site for analysis. The period of time between transmissions can vary, but one test used 30-minute intervals while on the high seas and that interval seems to be very useful. Each of the systems can also be pulsed to report its contacts upon request. NRL's MARS can actually transmit a continuous feed to the intended analysis center via Iridium satellites. Further analysis is needed to establish the correct interval. Efforts are underway to evaluate this capability on government and commercial vessels.

Coincident collection of AIS data with radar tracks, be they from space, or ships or ground or air-based, involving the efforts of the Coast Guard, Navy, and Customes and Border Protection would allow for both space and OTH radar systems and, in many cases, the several SARSats, to calibrate their sensors by providing ground truth on the position, size, course, and speed of the images they are currently collecting. Having a sufficient amount of this type of data would allow engineers to develop algorithms to extrapolate the findings to other cases. A joint offshore test concept development meeting was held at CSTARS in July 2006 to examine how to implement this concept, but funding was never secured. Hopefully future efforts will be more successful.

The future

The next system under discussion is a bit further away from fruition, if it ever gets there at all. Several years ago, NASA engineers placed a passive coherent location (PCL) receiver/ processor system in a business jet to see if they could use the energy transmitted down from several classes of spacecraft, including the transmissions of the global positioning satellite and international maritime communications satellites, reflected off the ocean to detect wave weights and currents. The tests were successful and some of those engineers believe those same transmissions could be used to detect ships, if a large enough antenna (100 meters?) could be lofted to 60,000 feet or more.

US DOD's Defense Advanced Research Project Agency (DARPA) is looking at developing just such an antenna to be placed on/in the skin of the high altitude airship, and similar craft in a project called Integrated Sensor Is Structure (ISIS). Indeed, one of the limiting factors of using such craft for maritime surveillance is the large size, weight and power (SWAP) requirements to place an air and/or maritime surveillance radar on board that would be capable of capitalizing on the high altitude, and its commensurate long line of sight. Using satellite transmission based PCL techniques as just described would mean there would be no need to carry a large radar on a high altitude airship used for maritime patrol. This concept is being discussed, but no additional tests have been run to date. Hopefully, this concept will be investigated further.

Other technologies being considered for the approaches zone (that area extending from beyond line of sight to approximately 100 miles offshore) include high-altitude, long-endurance unmanned air vehicles, such as a modified Global Hawk optimized to operate in a maritime surveillance role; medium-altitude, long endurance unmanned air vehicles such as the Predator-B/Mariner; and airships in a variety of configurations, including hybrids and unmanned versions. Also under consideration: aerostats capable of being launched from vessels underway and capable of remaining on station during all-weather except hurricanes; buoys equipped with a host of sensors, including AIS, surface wave radars, signals intelligence systems, and remote-control cameras; and remotely piloted/unmanned surface and subsurface vessels.

No one system is seen as being able to do it all, but a judicious mix of the above systems should allow the partner nations to detect, identify, track, and interdict nearly all vessels that approach its coasts. Indeed, there is no silver bullet, but there are some effective copper and silicon ones, and space systems are the key. We just need to collaborate on an international scale in order to realize their potential.

The international communities in both the maritime and space segments of the world already cooperate with each other in many ways. The International Space Station is a major case in point; another is the IALA and the IMO. The list is really quite long. This author proposes the international maritime and space communities take the next step and join in setting in motion the process needed to both build broad agreement, and the establishment of an international body to manage the process to build and run a Global Space Partnership, with their first order of business the implementation of the Collaboration in Space for International Global Maritime Awareness (C-SIGMA) concept.

It is early 2010 as I write this. The attempted terrorist attack of Christmas, 2009 has just happened and many people are in full flap over yet another attempt to destroy a jetliner, this time with 289 souls onboard. I have listened carefully to the many pundits and politicians and I have not heard one word about the vulnerability of our maritime frontier. Everyone is busy fighting the last war. I hope I, and many of my professional colleagues, are wrong, and we never have to learn exactly how damaging a coordinated attack on or via our maritime domain could be. I, and several of my colleagues, are actually a bit surprised such an attack has not already happened.

One thing has stuck with me in the past eight years. Many more Canadians than Americans get the seriousness of the threat. This is especially true of the members of the United States Navy, of which I was closely associated for many years. Not sure why this is true, but in my personal experience, it is clearly true.

What follows details my personal path to create an organization as well as develop and apply the technology to allow the United States and its partners world-wide to counter this threat.

Appendix Five
Space-based Global Maritime Awareness
Is a House

This article was published in March 2020
at spacewatch.global

As space-based Global Maritime Awareness (GMA) has become more and more robust, I have come to think of it as a house, even my home. The latest edition to this house, provided by the new Radio Frequency Geolocation satellites, have finally made this house truly livable.

Let me explain. Space-based Global Maritime Awareness (GMA) came into being with the launch of the first S-AIS constellation by ORBCOMM in 2008. It really was not complete as a system until the launch of unclassified radio frequency (RF) satellites. HawkEye 360's Pathfinder is probably the first RF geolocation satellite. Certainly, the first anyone is talking about. It has just completed a year in space, living up to all expectations. It will be the first of many. The term describing the new satellites is a bit cumbersome. Indeed, I originally called them unclassified ELINT (electronic intelligence) satellites but some folks were uncomfortable with that label. We know Soviet ELINT satellites were discussed and described in many open sources starting in the 1980s. and that is what these satellites are. Just to be politically correct, I suggest we call them RFgeoSats, but I am open to suggestions. You saw it here, first.

It has been recognized for some years that RF geolocation would be a very useful tool for maritime awareness, especially when used in collaboration with S-AIS. It fills a need to track ships when they turn off their AIS, as many bad actors now do when they commence nefarious actions such as smuggling or illegal fishing or pumping their bilges in restricted waters. But you still need AIS to identify the ships on initial contact, before they turn it off. These two systems, S-AIS and RFgeoSats are not in competition with each other as some suggest, but rather, they are complementary, maybe even synergistic. We do not have enough firm data to be absolutely sure, but initial indications point to it.

Back to the house analogy: S-AIS gives you the base, the foundation locational information for global maritime awareness. It does not give just location, but also name, physical dimensions, last port of call, next port of call, and a great deal more information. AIS is truly the sturdy foundation of Global Maritime Awareness, both terrestrial and space-based. RF Geolocation satellites are the floor set on that foundation. Heretofore, we had a foundation, but we needed more. We needed a floor to set on top of that foundation. That base foundation had its rough spots and could be pretty cold without more data and information to warm it up. The RFgeoSats collection and location of maritime emitters does just that. They upgrade the processing of these signals. When I was an ELINT operator on a cruiser in Vietnam, we called this "fingerprinting." This would allow us to determine exactly which ship, by name, we were picking up before we could see it coming over the horizon. The identification fingerprinting process became too technically sophisticated to call it simple fingerprinting so they came up with a much more important sounding, even elegant, term: Specific Emitter Identification (SEI). Not sure what the folks at Ft. Meade call it today, but the end result is the same. I have faith that we will soon, in many cases, be able to determine which specific ship is broadcasting a particular radar or communications signal by the uniqueness of its signal parameters. That information will be the result of the collaboration of S-AIS and RFgeoSats systems. These two systems will be clearly synergistic at that point.

To continue the house analogy, once the foundation is laid and the floor is in place you can start erecting the walls. This provides maritime awareness and thus security. Indeed, it is now widely acknowledged you cannot have maritime security without maritime awareness, and you cannot have effective maritime awareness without space systems being an integral part. The walls, a very important part of any structure, are provided by the synthetic aperture radar satellites. These are SARsats, with their day/night rain or shine capabilities. Early versions were big guys. I stood next to MDA's RadarSat 2 just before it was packed up and sent to the launch pad. I was truly impressed by its size... about 47 feet (15 meters) as I recall. I am told ICEYE, one of the newest SARsats, and almost as capable as the previous generation systems, could fit in the front passenger seat of a small car.

The pattern of life at sea or operations at sea, or call it what you will, gleaned from the collection of S-AIS for the past dozen years plus, and soon to be assisted by the RF geoStats, greatly assists the SARs to determine where to focus their collection efforts. By comparing the data collected by these three systems, analysts can collect a great deal of information including who is trying to avoid detection. The geographic area and the history of the vessel trying to avoid detection tells the analysts, now often assisted by AI and machine learning, a great deal. These three systems, S-AIS, RFgeoSats and SARsats, are a second true synergy. They complement each other when used in collaboration. Once the analysts have a handle on their task at hand, they may want to call for the final collection effort, the imaging satellites.

The optical satellites systems, still and video, are the windows of the GMA house. As when building a real house, once the walls are in place you can start adding windows. Windows are especially useful to look at specific objects and events. They have unique capabilities to tell the analysts a great deal, but you have to get the look in daylight at this time to really get the high-resolution pictures often required by analysts. It is an apt analogy as these are the systems that we use to see the world as it really is. But just like windows, they don't work very well when the lights are out, or even just dim. Still, they are very, very useful under the right circumstances. While imaging satellites can be used to search in known areas of maritime operations, they are most productive when they are aimed at a specific location for a specific purpose. The analysis to provide that location and purpose comes from the other three systems, S-AIS, RFgeoSats and SARsats, described above.

Another type of unclassified satellite system which heretofore was used for other purposes has recently started being pioneered by Global Fishing Watch to provide yet another data source for global maritime awareness. It is called Visible Infrared Imaging

Radiometer Suite (VIIRS). It is limited to night-time and fair weather as it detects ships illuminated at night. In that many fishing boats operate at night it is especially useful in the detection of illegal, unregulated fishing (IUU), a major problem worldwide.

VIIRS is very similar to SAR in its widest mode. Its large swath capacity provides no detailed discrimination but even a single pixel of light indicates the presence of at least one ship.

Even better, the data is free as the sensor is on both NOAA weather satellites as well as a NASA research satellite which covers the entire world several times a day. Altogether, it is proving to be another useful complement to S-AIS as well as SAR. I am sure the RF geolocation satellites data will also be added to this mix very soon, if not already, helping to identify at least the type of activity.

I have heard this new use of a nearly 10-year old system as an old dog in its kennel, providing watchdog functions, weather and time permitting. I prefer to call this new capability a skylight, or even more correctly a moonroof, for our house. Very useful when the time is right, but only at night. Perfect for observing night-time activities such as illegal fishing.

The Global Maritime Awareness house is decorated with many types of information in all sizes and shapes from sources including the IMO's Long Range Identification (LRIT) and the Vessel Monitoring System (VMS) of the world's fishing fleets, shipping records, police and other law enforcement records, ship builders, brokers and financial records. It is a very eclectic collection, but nearly all of it is useful at one time or another.

The roof is dual purpose. It both collects and protects the house and its decorations of stacks of the information available on maritime operations both at sea and in the marine support system on land. This roof is also made up of many different Dynamic Data Analysis (DDA) tools, now routinely incorporating machine learning and AI. The DDA tools are being supplied by many different entities, including, but not limited to, E-GEOS's Smart Eyes on the Seas (SEonSE), CLS's Maritime Awareness System (MAS), AirBus and MDA.

However, there are also stand-alone DDA efforts out there. These dynamic data analysis tools are, just like a roof in a real house, a crucial part of the structure. They are needed to take the derived and accumulated data and develop information, understanding and finally, wisdom, from the data collected.

The DDA tools have come a long way from Channel Logistics' Computer Assisted Threat Evaluation (CATE) tool of 2004 and Greenline's Computer Analysis Maritime Threat Evaluation System (CAMTES) of about two years later. Jatin Bains of Channel Logistics and his competitor, Paul Kerstanski of Greenline, were the pioneers in this field and to this day they have both continued to upgrade their tools' capabilities. As an example, Channel Logistics, is now doing business as Space-Eyes, and is revolutionizing collections by co-locating SAR and AIS payloads on the same spacecraft with a 400 km swath. The small satellite revolution will enable Tactical ISR collections over the maritime domain, however, analytics and insight at the speed of relevance is still the holy grail.

Windward, a Tel-Aviv based maritime analytics company headed by Ami Daniel, took DDA to a whole new level by being the first that I know of to fully introduce AI to exploit maritime data in innovative ways. They are providing insights to many organizations in the maritime ecosystem, including governments, insurers, financial institutions and energy companies, enabling them to optimize their performance and stay ahead of the adversaries.

Similarly, David Waldrop developed a DDA tool for the classified elements of the US government but now has permission to sell some parts of it in the commercial world. I have seen his brief and it is very impressive, but I have no real idea who has the best system today. Since I no longer work for the US government, I do not open access to the inner workings of these tools. I have been involved in the computer fusion and analysis of both like and dissimilar data since the early 1980s, and I can tell all four of these men are very bright and very dedicated to their craft. I am just smart enough to understand how smart they are.

Summary

This is the house I have been helping to build since October 2001, and I am very proud of the space-based Global Maritime Awareness residence. Indeed, the world space community keeps improving with more robust and capable systems. The foundation is pretty-well set, with three major players, ORBCOMM, exactEarth/Harris, and Spire competing with each other to build better and better S-AIS based databases and support feeds. The walls are looking better and better as MDA, DLR/AirBus and E-GEOS came on-line and then others, most notably ICEYE and SSTL's NovaSAR, and soon Capella, of which big things are projected. These SARs, from all over the Earth, are becoming more and more useful in the maritime domain.

The number of optical systems in orbit today, led by Planet Labs and Maxar, is simply mind boggling. This house now has many windows and doors with more being installed on a seemingly monthly basis.

The floor always needed improvement. Now that Hawkeye 360 is up and operating and KLEOS and Horizon Technologies, plus others, are ready to join, the flooring is getting stronger and better looking. I predict that the RFgeoSat floor, which is basically unknown in many parts of the world today, will soon have a shine, just like a new floor does. That is exactly what happened to S-AIS. In the space of two years, 2008-2010, S-AIS went from unknown to must have. I am sure the same thing will happen to the RFgeoSats.

I have been part of the global team building the Global Maritime Awareness house for over 18 years and I can see that it will soon be a mansion. I am also sure that if I were to come back 18 years from now, I would not recognize it. Just as Orville and Wilbur Wright would not recognize a 747, an A-380 or a F-35 if they were to come back today. Just like them, I had no idea exactly what I was helping create. The Wright brothers and I both saw a need and dedicated our lives to making our dreams happen. It is a great feeling.

If you think this was written to be a primer for space-based Global Maritime Awareness, you are exactly right.

Appendix Six
Space-based Global Maritime Awareness
Is About to Come of Age

Space-based Global Maritime Awareness (GMA) is finally, after 19 years, on the edge of coming of age. All of the efforts of the past 19 years to bring GMA into the world as a useful entity globally are showing signs of bearing fruit at last. The unique capabilities provided by the new Radio Frequency Geolocation satellites to GMA, are quickly going from concept to early maturity. This capability may well be the tipping point for GMA.

GMA was conceived in 2001 as a means to combat maritime terrorism and came into being with the launch of the first S-AIS constellation by ORBCOMM in 2008 but it really was not complete as a system until the recent launch of the first unclassified radio frequency (RF) satellites in the last eighteen months. HawkEye 360's Pathfinder was the first RF geolocation satellite, quickly followed by Unseen Labs BRO-1. Both systems have now completed a year in space, living up to all expectations. I predict these will be the first of many. Maybe not as many as the +/- 175 S-AIS satellites now in orbit, but certainly there will be several dozen, with near continuous global coverage.

The term describing the new satellites is a bit cumbersome. Indeed, I originally called them unclassified electronic intelligence (ELINT) satellites but some owners and builders of these new satellites are uncomfortable with that label even though Soviet ELINT satellites have been discussed and described in many open sources starting in the early 1980s and that is what these first RF geolocation satellites bring to mind.

I do understand why many people in the intelligence world are uncomfortable with calling them that. In the early days you could not say *intelligence* and *satellite* in the same sentence. The people with that memory will probably be even more uncomfortable with Amber, the new satellite by Horizon Technology. According to the company, it has the capability to collect and exploit unencrypted communications as well as radars and other emitters, thus adding an unclassified Communications Intelligence (COMINT) capability to the space world within the year.

The combination of both unclassified ELINT and COMINT gives us a true Signals Intelligence (SIGINT) capability in space that we can share across the globe for the first time ever. It is that fact that really raises my expectations. Back in 2004 the intelligence organizations of both the United States and Canada tried and failed to get satellite AIS (S-AIS) declared a classified SIGINT system while ORBCOMM was building the first S-AIS constellation. I know this because I was working at the multi-agency Maritime Domain Awareness Program Integration Office (MDA/PIO), the organization which funded the first S-AIS constellation, and I was tasked with writing the rebuttal.

The Secretary of Defense, the boss of the intelligence agency which was leading the attack on S-AIS asked for the two parties, the MDA/PIO and the National Security Agency (NSA), to write a four or less page paper outlining their respective positions, giving the benefits, pros and cons of why or why not S-AIS should be classified. My paper was less than three, the NSA's was four full pages with narrow margins. The apprehension as we waited for the decision was not much fun, but common sense did win out and S-AIS was declared to be an unclassified Aid To Navigation (ATON) and an unclassified global maritime situational awareness tool as originally intended. But, with that experience behind me, I am more than a bit amazed these new systems are being permitted, but very clearly times have changed. With these thoughts in mind, and just to be politically correct, I suggest we call these new satellites Radio Frequency Geolocation Satellites, RFgeoSats for short, or maybe RFGSats, but I am open to suggestions as I admit the title itself is a bit cumbersome.

It has been recognized for some years that RF geolocation would be a most useful tool for maritime awareness, especially when used in collaboration with S-AIS and synthetic aperture radar satellites (SARSats). The RFgeoSats fill a need to track ships when they turn off their AIS, as many bad actors now do when they commence nefarious actions such as smuggling or illegal fishing or pumping their bilges in restricted waters. It is especially true for sanction avoiders. But you still can use AIS and S-AIS to identify the ships on initial contact before they turn it off. These two systems, S-AIS and RFgeoSats are not in competition with each other as some suggest, but rather, they are very complementary, probably even synergistic. We do not have enough firm data to be absolutely sure, but initial indications point to it.

S-AIS gives you the skeleton on which to build GMA, not just space-based GMA, but true Global Maritime Awareness. It is the foundation of GMA by providing locational information for all legal ships engaged in international commerce plus many more that have installed AIS as a safety device, the use for which it was originally built. It does not give just location, but also name, physical dimensions, last port of call, next port of call, and a great deal more information. AIS, both terrestrial and space-based, is truly the sturdy skeleton of Global Maritime Awareness. Heretofore, we had that skeleton, but needed muscles to move the skeleton forward, and the RFgeoSats lends capabilities for collection, location and identification of emitters.

All new systems routinely need refinement and the RFgeoSats are no different as they are just commencing operations. It is natural to believe the initial systems will need upgrades to both their processing and reporting abilities. One ability which will need special attention is the ability to do Specific Emitter Identification. When I was an ELINT operator on a cruiser off Vietnam in 1968 we could not do that, but when I came back in 1972 we could. We called this fingerprinting. It allowed us to determine exactly which specific emitter we were collecting, by name or geolocation. If it was a ship-borne emitter we were picking up, we could know its name before we could see it coming over the horizon. The fingerprinting process became too technically sophisticated to call it simple fingerprinting so they came up with a much more important sounding, even elegant, term: Specific Emitter Identification (SEI). Not sure what the folks in the Intelligence Community call it today, but the end result is the same. I am told at least three of the four companies now building and operating RFgeoSats either have, or are working on, this capability and I have faith that the RFgeoSats will soon be able to determine which specific ship is broadcasting a particular radar or communications signal by the uniqueness of its signal parameters.

That information will be the result of both the RF engineers working for these companies and others learning how to use S-AIS and RFgeoSats systems in collaboration. These two systems will be clearly synergistic at that point.

Once the skeleton and muscles are developed the brain needs to be developed and matured as well. A body with its skeleton and muscles will be powerful but it is limited in its utility without a brain. It needs more data and information to breathe life into it and give it life, and to warm it up and give it direction and purpose. Dynamic data processing provides the ability to convert the data collected into information. Further processing creates understanding and, hopefully, develops wisdom which becomes maritime awareness and thus security. Indeed, it is now widely acknowledged you cannot have maritime security without maritime awareness, and you cannot have effective maritime awareness without space systems being an integral part.

Synthetic aperture radar satellites (SARsats), with their day/ night, rain or shine capabilities, are the eyes of the GMA system. Early versions were large. In the Fall of 2007 I stood next to MDA's RadarSat 2 just before it was packed up and sent to the launch pad. I was truly impressed by its size, a width of more than 15 meters. I am told ICEYE, one of the newest SARsats, and almost as capable as the previous generation systems, is less than a quarter that size. Indeed, its prototype was tested by flying it in the front passenger seat of a Cessna 172, a small single engine airplane. SARsats and S-AIS have proven to be synergistic. Both are good, useful systems by themselves, but they are great together. I anticipate the RFgeoSats will join this duo and further improve space-based global maritime awareness by a significant factor.

The pattern of life at sea or operations at sea, or call it what you will, gleaned from the collection of S-AIS for the past dozen years plus, and soon to be assisted by the RFgeoSats, greatly assists the SARs to determine where to focus their collection efforts. By comparing the data collected by these three systems, analysts can develop a great deal of information including who is trying to avoid detection. These three systems, S-AIS, RFgeoSats and SARsats, will be a second true synergy. They complement each other when used in collaboration. The geographic area and the history of the vessel trying to avoid detection tells the analysts, now often assisted by AI and machine learning, a great deal. Once the analysts have a handle on their task at hand, they may want to call for the final collection effort, the imaging satellites.

The optical satellites systems, still and video, are also part of the visual system of the GMA system. As when a young person matures they come to understand what they are seeing. This is especially useful to look at specific objects and events of things you are interested in seeing. Eyes have unique capabilities to tell the analysts a great deal, but you have to get the look in daylight at this time to really get the high-resolution pictures which are most useful to analysts. It is an apt analogy as these are the systems that we use to see the world as it really is. But just like eyes, they don't work very well when the lights are out, or even just dim. Still, they are very, very useful under the right circumstances. While imaging satellites can be used to search in known areas of maritime operations, they are most productive when they are aimed at a specific location for a specific purpose. The analysis to provide that location and purpose comes from the other three systems, S-AIS, RFgeoSats and SARsats, described above.

Another type of unclassified satellite system which heretofore was used for other purposes has recently started being pioneered by Global Fishing Watch to provide yet another data source for global maritime awareness. It is called Visible Infrared Imaging Radiometer Suite (VIIRS). It is limited to night-time and fair weather as it detects ships illuminated at night. In that many fishing boats operate at night it is especially useful in the detection of illegal, unregulated fishing (IUU), a major problem worldwide. VIIRS is very similar to SAR in its widest mode. Its large swath capacity provides no detailed discrimination but even a single pixel of light indicates the presence of at least one ship.

Even better, the data is free as the sensor is on both NOAA weather satellites as well as a NASA research satellite which covers the entire world several times a day. Altogether, it is proving to be another useful complement to S-AIS as well as SAR. I am sure the RF geolocation satellites data will also be added to this mix very soon, if not already, helping to identify at least the type of activity. VIIRS is especially useful for observing night-time activities such as illegal fishing.

Global Maritime Awareness now has a great deal of information available to it in all sizes and shapes from multiple sources including the IMO's Long Range Identification and Tracking (LRIT) and the Vessel Monitoring System (VMS) of the world's fishing fleets, shipping records, police and other law enforcement records, ship builders, brokers and financial records. It is a very eclectic collection, but nearly all of it is useful at one time or another.

The Dynamic Data Analysis (DDA) system supporting GMA is dual purpose. It both collects and stores the data. We have been collecting more and more data and information on maritime operations both at sea and in the marine support system on land. The GMA system contains many different DDA tools, developed by entities all over the world. A commonality is the movement to routinely incorporate machine learning and AI. DDA tools have been developed by many of the companies which build remote sensing satellites including E-GEOS's Smart Eyes on the Seas (SEonSE) and the CLS's Maritime Awareness System (MAS). AirBus and MDA also have similar systems.

However, there are also stand-alone DDA efforts out there. The DDA tools have come a long way from Channel Logistics' Computer Assisted Threat Evaluation (CATE) tool of 2004 and Greenline's Computer Analysis Maritime Threat Evaluation System (CAMTES) of about two years later. Jatin Bains of Channel Logistics and his competitor, Paul Kerstanski of Greenline, were the pioneers in this field and to this day they have both continued to upgrade their tools' capabilities. Channel Logistics, now doing business as Space-Eyes, recently hired Paul thus bringing two of the original pioneers in the DDA field together as collaborators rather than competitors. I expect even greater things from this dynamic duo.

The satellite revolution now underway will enable significantly improved tactical ISR collection over the maritime domain, however, analytical insight at the speed of relevance is still the holy grail. Windward, a Tel-Aviv based maritime analytics company headed by Ami Daniel, took DDA to a whole new level by being the first that I know of to fully introduce AI to exploit maritime data in innovative ways. They are providing insights to many organizations in the maritime ecosystem, including governments, insurers, financial institutions and energy companies, enabling them to optimize their performance and stay ahead of the bad actors they are hunting.

Similarly, David Waldrop developed a DDA tool for the classified elements of the US government but now has permission to sell some parts of it in the commercial world. I have seen his brief and it is very impressive, but I have no real idea who has the best system today. Since I no longer work for the US government, I do not open access to the inner workings of these tools.

I have been involved in the computer fusion and analysis of both like and dissimilar data since the early 1980s. All four of these men are very bright and very dedicated to their craft. I am just barely smart enough to understand how smart they truly are.

The way forward

Having been directly involved in building GMA since October 2001, I now believe GMA is right on the edge of fulfilling the promise we saw back then. Indeed, the world space community keeps upgrading the GMA system with both more satellites with improved capabilities. Its skeleton is pretty-well formed, with three major players, ORBCOMM, exact Earth/Harris, and Spire competing with each other to build better and better S-AIS based databases and support feeds. Its eyes are getting better and better as MDA, DLR/AirBus and E-GEOS came on-line and then others, most notably ICEYE and SSTL's NovaSAR. Capella, of which big things are projected, has just joined this group. These SARs, from all over the Earth, are becoming more and more useful in the maritime domain.

The number of optical systems in orbit today, led by Planet Labs and Maxar is simply mind boggling. This house now has many windows and doors with more being installed on a seemingly monthly basis.

The muscles, good enough now, will continue to improve. Now that Hawkeye 360 and Unseen Labs are up and operating, with KLEOS and Horizon Technologies ready to join, the muscles are getting stronger. I predict that the RFgeoSat system, which is basically unknown in many parts of the world today, will soon have a major role in the GMA system.

That is exactly what happened to S-AIS. In the space of two years, 2008-2010, S-AIS went from unknown to must have. I am sure the same thing will happen to the RFgeoSats. We are now at 2008 as far as the utility of RFgeoSats. Their full utility has only just begun to be explored. If history is a guide, the possibilities are exceptional, the only limitation is the imagination of the developers. That was certainly true of S-AIS. It was the users that keep coming up with new ways to exploit the information provided to this day, twelve years later. Much the same as satellite navigation. Few people not associated with Johns Hopkins' Applied Physics Laboratory (JHU/APL) recall satellite navigation was originally conceived and implemented to provide launch point accuracy for submarine launched ballistic missiles. Now its child, GPS, is ubiquitous. The same thing has happened to S-AIS as far as the maritime world. It, too, was conceived at JHU/APL.

Regarding RFGeoSats, one of the first things that needs to be done is the development of an Electronic Support Library (ESL) which contains the parameters of every shipborne emitter which has ever been collected. This is a huge task, but I believe it is completely doable given the big-data collection and analysis capabilities of the 21st century.

We now have multiple entities collecting and processing S-AIS and AIS historical data going back to at least 2002, two years before AIS became mandatory. Big data has continued to expand rapidly in the last 20 years and our ability to store, access and analyze large amounts of data will continue to expand. The parameters of an emitter are a finite set that can be stored in a database with a set number of fields. Indeed, we are probably looking at a data base with something less than 100 fields, even including the physical dimensions of the specific platform the emitter with which it is associated, plus the identifying characteristics of the signal itself, plus any pertinent operational and financial history.

Summary

I have been part of the global team raising this system called Global Maritime Awareness for 19 years now, and I can see that it will be a champion soon. I am also sure that if I were to come back 19 years from now, I would not recognize it. Just as Orville and Wilbur Wright would not recognize a Boeing 747, an Airbus A-380 or a Lockheed F-35 if they were to come back today. Just like them, I had no idea exactly what I was helping create. The Wright brothers and I both saw a need and dedicated our lives to making our dreams happen. It is a great feeling.

Glossary

Initialisms, Acronyms, Vernacular of the Realm

	Pertinent to all books, articles and papers written by this author
ACTD	Advanced Concept Technology Demonstration, forerunner to JCTDs. ACTDs did not have an interservice, much less inter-service, focus.
AIS	Automatic Identification System (AIS), a UN-created collision avoidance line-of-sight beacon and traffic control operating in the VHF frequency band. Range of about 25 miles.
AMIC	Air and Marine Interdiction Center, forerunner to the AMOC
AMOC	Air and Marine Operations Center. A US Customs and Border Patrol (CBP) anti-smuggling center in Riverside California which monitors all Western Hemisphere air traffic between Peru and the Arctic Circle
APL	Johns Hopkins University Applied Physics Lab
CAPT	Navy Captain (officer of the 6th rank (O-6); same pay as a colonel
CAR (NWC)	Center for Advanced Research (at the Naval War College). One part of the Center for Naval Warfare Studies (CNWS)

CG	Guided Missile Cruiser
C-SIGMA	Collaboration in Space for International Global Maritime Awareness
EC-121M	US Navy 4-engine turbo-propeller reconnaissance aircraft. First aircraft to use computers to run its communications intelligence (COMINT) collections system
Echo-Class Submarine	Soviet guided missile nuclear powered submarine (SSGN)
ELINT	Electronic Intelligence: The gathering and analysis of radar and other emitter signal characteristics.
EP-3B	US Navy 4 turbo-propped reconnaissance aircraft. Derivative of the P-3 antisubmarine aircraft especially configured for signals intelligence (SIGINT) collection.
EP-3E	US Navy 4-engine turbopropellor reconnaissance aircraft First acft to use computers to run its communications intelligence (COMINT) collections system.
GMA	Global Maritime Awareness, the concept of knowing what is happening in the maritime domain, globally. It subsumes both MDA and MSA.
Il-28	Soviet-built jet-engine light bomber

IMO	International Maritime Organization, a specialized agency of the UN dealing with regulating and protecting the maritime domain.
JCTD	Joint Capabilities Technology Demonstration (JCTD) The program to rapidly evaluate new technology for the Department of Defense. It is now open to all US government agencies.
JPO	Joint Program Office. A generic term referring to a program which pertains to more than one service.
LIME	The LIME assessment framework is a methodological tool to compare, in the context of the Lisbon Strategy, the performance of EU Member States in terms of GDP and in terms of 20 policy areas affecting growth.
LOE	Limited Objective Experiment
LRIT	Long Range Identification and Tracking, an IMO regulation requiring all SOLAS class ships to identify themselves and give their position, course and speed under several different requirements.
MARISA	Maritime Integrated Surveillance Awareness
MDA	Maritime Domain Awareness, a concept dealing with acquiring knowledge of what is happening in the maritime domain

MDA/PIO	Maritime Domain Awareness/Program Integration Office. Created in early 2003 to integrate all US maritime surveillance programs not focused on combat.
MiG-17	Soviet fighter first flown in the mid-1950s. Subsonic, but fast for its day.
MiG-19	Soviet fighter first flown in the late 1950s. First Soviet mass-produced supersonic aircraft.
MiG-21	Soviet fighter first flown in the 1960s. Later versions were capable of near Mach 2 speeds.
MiG-25	Soviet fighter first flown in the mid-1970s. Very high altitude with near Mach 3 speed. Very fast for its day.
MSA	Maritime Situational Awareness, similar to MDA but more focused at the tactical level.
MSSIS	Maritime Safety and Security System. A Department of Transportation program, initially funded by the MDA/PIO and subsequently adopted by DoD to track ships using AIS, including S-AIS via collaboration by many nations.
Naval Security Group NavSec-Gru	The service cryptologic organization of the US Navy. It had two bosses, the Navy and the National Security Agency. Disestablished in 2005 with missions shifted to Naval Network Warfare Command and US Tenth Fleet.

NavSecGruAct (NSGA)	Naval Security Group Activity: A Major NSG command, generally commanded by a Navy captain
NavSecGruDet (NSGD)	Naval Security Group Detachment (reports to a NSGA)
NMIIO	National Maritime Intelligence Integration Office, the national office created in 2007 to marry all US government classified maritime security efforts into one organization.
NSA	National Security Agency. The part of the US Intelligence Community charged with conducting Signals Intelligence (SIGINT). Its two main components are ELINT (electronic intelligence) and COMINT (communications intelligence).
NSC	National Security Council, the part of the US government executive branch charged with guiding the national security policies and interests of the USA.
NSC	National Security Council. The part of the Executive Branch which deals with security policy and associated guidance.
NWC	Naval War College, the senior education institution of the US Navy. It also has a robust research department known as the Center for Naval Warfare Studies, one element of which is Center for War Gaming, and another is the Strategic Studies Group.

OGMSA	Office of Global Maritime Situational Awareness, the national office created in 2007 to develop maritime situational awareness at the unclassified and law enforcement sensitive (LES) levels.
ONI	Office of Naval Intelligence. The lead intelligence agency focused on maritime intelligence in the USA.
ROTHR	Relocatable Over-the-Horizon Radar
R&D	Research and development, the multi-step process whereby new systems/products are created.
RC-135M (Rivet Card)	USAF 4 jet-engined reconnaissance aircraft based on the KC-135 tanker aircraft. Forerunner to the RC-135V and RC-135W.
RC-135W (Rivet Joint)	USAF 4 jet-engined reconnaissance aircraft, First USAF aircraft to use computers to run its communications intelligence (COMINT) collections system.
RC-135V/W (Rivet Joint)	Family of USAF 4 jet-engined reconnaissance aircraft based on the KC-135 tanker.
S-AIS	Satellite Automatic Identification System (AIS), a ship detection, identification and tracking system using AIS first used in 2008.

SAR	Synthetic Aperture Radar, a radar found on aircraft and space systems uses the motion of the radar antenna over a target region to provide finer spatial resolution than conventional stationary beam-scanning radars.
SOLAS	Safety of Life at Sea: An IMO-defined class of ships which must abide with a specific set of rules; aka SOLAS Class ship.
SSN	Nuclear powered attack submarine
T&E	Test and evaluation, the final phase of an acquisition program before a product/system is released for intended use.
Tu-16	Soviet-built medium bomber, some converted to reconnaissance roles, others to carry missiles.
Tu-95	Soviet-built long ranger bomber, some converted to a reconnaissance role, others to carry missiles.
VIP	Vessel Identification and Positioning, Volpe Center tool prior to AIS, switched to AIS as its basis.
WAL (& WALEX)	Warfare Analysis Lab at Johns Hopkins APL. A unique marriage of wargaming and system engineering to rigorously look at all issues, including technology, of the subject at hand. These events were called WAL Exercises (WALEXs).

Letters and Memoranda

Space Foundation Award letter January 6, 2021

Dear Mr. Thomas:

On behalf of the entire Space Foundation team, let me congratulate you and C-SIGMA, on your lifetime contribution to maritime safety using space system capabilities. Work such as yours demonstrates the ingenuity and commitment that comes fro and applying creativity and perseverance to be delivering solutions that make all the difference in the world. It is also testament to the reach and impact that our space programs have for shaping and improving life here on Earth.

Because of your work, commerce and supply chain operations can flow safely around the globe, while security threats and needed resilience to ever dynamic risks are adequately addressed by our nation and its allies. Your work in these areas with the Navy, Air Force, Coast Guard, Johns Hopkins APL, the Department of Homeland Security, the private sector and others is testament to the character and dedication you have brought to all of your duties. And as a retired Navy officer, I can personally attest my service career and of those of sailors worldwide have benefited from your works and lifetime of service.

As my staff shared your story with me, we wanted to recognize you, your work and the lifetime of difference it has made. Enclosed you will find a personal award from the Space Foundation to showcase and share with your partners and family as testament to a career that has not only enhanced breakthrough space technologies and applications, but shaped commerce, national security and public safety for the better.

We hope it will be a source of pride for you but also be an inspiration for others in your orbit to see a career of service that has transformed life and our world for the better. Work such as yours is more than impactful—it's pioneering and that is a legacy to celebrate.

Thank you again for your service and I hope you enjoy this token of our respect and admiration for you and your work.

Sincerely,
Tom Zelibor
Rear Admiral, USN (Ret.) CEO, Space Foundation

Johns Hopkins Memorandum
July 28 2020

Former Johns Hopkins Applied Physics Laboratory (APL) employee Guy Thomas knows a thing or two about persistence.

In response to a direct order issued on 9/11 by President George W. Bush to the Chief of Naval Operations, the Navy established a special presidential taskforce looking for ways to mitigate maritime terrorism, including ways to help the US identify and track all vessels approaching the country. Thomas—then an APL employee in the National Security Analysis Department and liaison to the Naval War College—served as the technology co-lead on the effort. At that time the existing Automatic Identification System (AIS) system worked for short-range detection, but long-range tracking wasn't considered possible.

A former US Navy space systems subspecialist, Thomas postured that a space-based solution was the answer to offshore tracking and identification and went looking for an answer via space systems. When he received a brief on AIS "I recall asking whether anyone ever thought about putting an AIS receiver on a low Earth orbiting satellite," said Thomas. The answer from the five engineers in the room was the concept was not feasible. Thomas thought they were wrong and set out to prove it, with the help of his APL colleagues.

It would take another three years to sell the idea and receive funding for a prototype, and then a four-year battle with a number of other organizations over many different items including with NSA which tried to classify the concept, but his years-long determination to form the Satellite Automatic Identification System (S-AIS) is now being recognized with a nomination for the National Medal of Technology and Innovation.

The prestigious NMTI award is the highest honor bestowed upon individuals for technological achievement in the US Presented by the president, the medal is awarded to innovators for their outstanding contributions to America's economic, environmental and social well-being. The purpose of the National Medal of Technology and Innovation is to recognize those who have made lasting contributions to America's competitiveness, standard of living, and quality of life through technological innovation.

"I'm very honored to receive this nomination for work that I started while at APL," said Thomas. "My network of APL colleagues were a critical element of support in both the development of the hypothesis, and the dogged pursuit which finally resulted in the creation of S-AIS. It probably would not have happened without them."

Thomas' work on S-AIS has helped transform maritime safety and security operations globally. No longer just a concept, S-AIS is now the primary ship identification and location system for vessels world-wide. Its information is widely used as the core of Maritime Domain Awareness, search and rescue, environmental monitoring, resource protection and maritime intelligence application such as counter-smuggling and sanctions violations detection.

But back in 2001, S-AIS was a hard sell.

Thomas faced hurdles at every turn. He had to convince stakeholders of the value of the system and that it could actually work. There were doubts that a signal could be picked up from space because of the density in that portion of the frequency spectrum. Some simply thought the idea was not worth the cost.

Thomas eventually secured enough funding through the US Coast Guard and negotiated a public-private partnership with ORBCOMM, a satellite communications company, to build an initial prototype. The first six commercial S-AIS satellites were launched in 2008. S-AIS has since helped the US Coast Guard and navies and coast guards all over the world observe approaching ships and vessels and has also become a primary tool for the protection of the maritime environment and its resources.

"[Thomas'] persistence in fostering collaboration and pursuit of S-AIS dual use for both commercial and government sectors resulted in a practical solution for maritime logistics and a national security layer against non-cooperative vessel threats," former US Deputy Undersecretary of Defense John Kubricky stated in a letter nominating Thomas for the award.

"Few technologies have had such a significant impact on the vitality of this nation's economy as has S-AIS, which will remain the dominant baseline for decades of maritime operations in the future."

National Medal of Technology Nomination Letters

In 2020, Guy Thomas was nominated for the National Medal of Technology and Innovation, the nation's highest honor for technological achievement, which is bestowed by the president of the United States. Beneath are letters supporting his nomination.

Nomination by Jeffrey P. High
(Former director, US Coast Guard Maritime Domain Awareness Program Integration Office)

Dear Committee Members,

I recommend George Guy Thomas for the National Medal of Technology and Innovation because of his extraordinary vision and persistence in bringing space-based Satellite Automatic Identification System (S-AIS) to the world, let alone the United States.

When I first met Guy in 2003, I was the first Director of the USCG Maritime Domain Awareness Program Integration Office (MDA-PIO), an SES-level position created in response to the terrorist attack on 9/11. Later, I hired Guy as my Science and Technology Advisor due to his background and knowledge of MDA (including his background in the intelligence community).

Before MDA-PIO, I was the USCG SES Director of Waterways Management, where one of my collateral duties was Head of the United States Delegation to the International Maritime Organization (IMO, an arm of the United Nations), Navigation Safety Committee. At the IMO, my US team joined other nations in support of the implementation of Automatic Identification System (AIS), an FM-based transponder system initially designed for collision avoidance.

After 9/11, I (and others in the USCG) recognized the potential value of AIS for security and Maritime Domain Awareness in general. ADM Loy, USCG Commandant at the time, spoke before the IMO and got the international community to advance the implementation of AIS.

During the early days in the MDA-PIO, I held morning staff meetings to assess our progress on many fronts in our efforts to improve MDA. I distinctly remember one meeting where we were brainstorming how to expand the use of AIS to improve our MDA.

Since it is essentially a line-of-sight system we talked about building a network of coastal towers to receive the signals (later this became Nationwide AIS, a large USCG program).

We also spoke about installing AIS receivers on the sea buoys at the entrance of all US harbors to push our visibility further offshore. Since those buoys generally have no power source, this idea morphed to looking into installing AIS on the NOAA offshore national buoy system (which USCG and NOAA pursued until a better idea came along).

The best idea came from Guy Thomas. He said we should put AIS receivers in satellites and watch from space. This, he argued, would give us an almost worldwide view of the ships carrying AIS. After the meeting, I asked Guy how we could confirm that his idea was possible and how we might implement it (at the time I did not know Guy had been thinking and talking about this for years).

Guy mentioned that ORBCOMM had a satellite system that he knew would be soon updating their satellites, so I asked him to set up a meeting between them and USCG. Guy also told me Johns Hopkins could do a feasibility study to show us that S-AIS could work and, specifically, that the collection of many similar signals could be deconflicted. Based on Guy's wise advice and counsel, I entered into a small feasibility study contract with Johns Hopkins on behalf of the USCG.

Later, after the successful feasibility study was completed, I devoted most of my budget to a single-source contract with ORBCOMM to carry a first test AIS receiver into space. That initial contract was expanded to include additional satellites and as we all know now, other providers have entered the market and S-AIS is alive and well across the world.

Benefits of S-AIS: The USCG's initial interest in S-AIS was for security. That was the primary motivation for the establishment of the MDA-PIO and our first objective. However, at an early Cabinet level meeting of many intelligence and security Departments, the participants at the four-star and higher level defined MDA as: "Maritime Domain Awareness is the effective understanding of anything associated with the maritime domain that could impact the security, safety, economy, or environment of the United States."

Clearly, security and safety are elements of "social well-being."

S-AIS has been one of the most effective tools for the enhancement of MDA and thus the "economic, environmental, and social well-being of the United States."

Without a doubt, the safe and secure international flow of passengers and cargo around the world is essential to our economy and our social well-being.

Likewise, our environment is much better protected from potential pollution, overfishing, etc. due the knowledge that there are eyes in the sky to detect bad actors. There is one important perspective I can share about economic benefits of MDA and S-AIS:

In my job as Director of Waterways Management, I led an interagency team that prepared a report for Congress on the US Maritime Transportation System (MTS). Essentially our report, signed by the Secretary of Transportation in 1999, explained the value of cargo (billions of dollars) that flowed through the US MTS. The report also showed that any disruption to the flow of commerce would be catastrophic, for both the US and the world economies. Therefore any climatic or man-made event (like terrorism) that could close a port could be devastating.

As a result, security of our ports and other MTS components was a major goal of the report to Congress. This is the direct tie to MDA (as defined), one of the reasons I was reassigned to start the MDA-PIO and ultimately the value of MDA and the S-AIS tool.

As of this writing, MDA, in large part supported by S-AIS, has been successful in avoiding any terrorist activity that would have affected our maritime-based economy. While the benefits of S-AIS clearly leverage America's competitiveness, standard of living and quality of life, it also has provided similar benefits to the global community. As they say, a rising tide raises all boats, a nautical reference that is most appropriate for this space-based technological innovation called S-AIS.

In closing, I offer my strongest recommendation for your consideration of Guy Thomas' vision and pursuit of the S-AIS dream that has turned out to exceed even my (if not his) wildest imagination of the potential possibilities when we set out to implement S-AIS.

—Jeffrey P. High (Former director, US Coast Guard Maritime Domain Awareness Program Integration Office)

Nomination by John J. Kubricky
(Retired, Deputy Under Secretary of Defense– Advanced Systems and Concepts)

Dear Committee Members:

I nominate George Guy Thomas for the National Medal of Technology and Innovation (NMTI) for his contributions to maritime safety, global logistics and international security demonstrated by his development of advanced signaling techniques and space-borne systems to identify, locate and track vessels throughout the world.

Professor Thomas's invention and implementation of the Satellite-Automatic Identification System (S-AIS) has become a pivotal enabler of our Nation's economic leadership, safety and security in global maritime operations and management.

As Director of Department of Homeland Security's (DHS) Systems Engineering and Advanced Research Projects Agency, my highest priorities included developing, testing and rapidly fielding safe, secure and efficient supply chain management systems that would ensure the uninterrupted flow of goods and services to benefit citizens of the United States and our Nation's economic infrastructure. I became familiar with the concept and development work of George G. Thomas on S-AIS while he and his USCG colleagues were seeking DHS funds to advance the project. It was clear in my first briefing that S-AIS was a game-changing concept and near-term technology solution that could be integrated, demonstrated and affordably deployed to meet DHS maritime security and economic challenges on an accelerated schedule.

It was my job to ensure that S-AIS was more than just a good idea to become a vital component of our Nation's maritime security; the successful S-AIS project needed to have the right people, the right resources and the right schedule to attract high-level support from DHS and Department of Defense (DoD) leadership, to include cost sharing and project collaboration. George Thomas had made-to-order skills, knowledge and experience, and he was able to garner support from his USCG leadership, as well as industry collaboration and acceptance— all elements necessary for success of this unique technology to become a ubiquitous system. In addition, George Thomas drove his invention to an elegant implementation by being affordable and useful for the maritime vessel operators who would have to adopt S-AIS for it to become a universal maritime system.

It should be mentioned that AIS was not a new technology at this time. In fact, AIS was already in use by the maritime industry. Having built a fleet of advanced surveillance aircraft and three research ships during my 34 years of industry experience before joining the federal government.

I knew terrestrial-based AIS serviced only maritime vessels that were within line-of-sight towers. As soon as I learned of George Thomas's concept and proposed plan, I knew his Satellite-based AIS was a disruptive technology that would change how the world manages vessel operations.

While my project teams at DHS Science & Technology supported the US Coast Guard's S-AIS concept, advanced development and demonstration phases in 2003 through 2005, we noted the continuous progress and achievements made under George Thomas's S-AIS project leadership.

All project milestones were successfully met or exceeded, which put this emerging capability on a fast track for broad implementation. At the conclusion of my third year at DHS, I returned to DoD as its Deputy Under Secretary for Advanced Systems. It was easy for me to select S-AIS as a crucial emerging capability that required interagency resources to flourish into full deployment and ultimately, worldwide commercialization for our Nation's global maritime trade partners.

It was during this period of 2006 to 2009 that George G. Thomas continued the relentless pursuit of team collaboration among government and industry participants in a broader use of S-AIS. Professor Thomas's participation at DoD events on maritime operations and logistics resulted in unique teamwork between disparate entities with conflicting or dissimilar missions. He would identify and drive to achieve objectives that yielded shared success among organizations. Guy's ability to form alliances was a crucial component in the successful advancement of S-AIS to full acceptance of the new technology's utility, practicality and commercialization.

In addition to the commercial utility of S-AIS in maritime operations, George Guy Thomas has also enabled a national security advantage for the United States in Maritime Domain Awareness. Guy's persistence in fostering collaboration and pursuit of S-AIS dual use for both commercial and government sectors resulted in a practical solution for maritime logistics AND a national security layer against non-cooperative vessel threats. Those who threaten the United States via maritime venues are now visible to federal, state, industry and non-governmental organizations as they approach our maritime borders and waters. The essence of S-AIS is one of full-spectrum surveillance that provides speedy safety and security to America's citizens, ships and waterways.

It is an honor to nominate George Guy Thomas for his efforts in conceiving, advancing and developing the S-AIS capability that provides the United States and its global trading partners with highly effective solutions to both maritime commerce and maritime security. Guy's method of developing S-AIS as an unclassified system that is available to commercial organizations is largely responsible for its positive contributions to our Nation's leadership in maritime logistics.

Few technologies have had such a significant impact on the vitality of this Nation's economy as has S-AIS, which will remain the dominant baseline for decades of maritime operations in the future.

—John J. Kubricky (Retired, Deputy Under Secretary of Defense –Advanced Systems and Concepts)

Nomination by Julio J. Gutiérrez
(Capt., US Navy (Ret.); Maritime Security Consultant)

Dear Committee Members:

I nominate George Guy Thomas for the National Medal of Technology and Innovation (NMTI) for his unique, original contributions to US and international maritime safety, trade, and security by his innovation of the Satellite-Automatic Identification System (S-AIS) spaceborne systems, methods, and architectures to locate, track, identify, and avoid collisions of vessels worldwide. His further innovative coalescing of hundreds of international corporations and other private users in the Collaboration in Space for Integration of Global Maritime Awareness (C-SIGMA) Center and umbrella organization he created since 2005 has facilitated the revolutionary rapid explosion of global-coverage multi-sensor affordable commercial space technologies and data services—and is also deserving of the NMTI.

Guy Thomas's creation and implementation of S-AIS and C-SIGMA critically support American and global maritime security, safety, economic progress for US$ trillions in maritime trade, and environmental stewardship of the oceans. I have known Guy Thomas as a professional colleague, mentor in advanced technologies, and as a friend since 1999.

For over two decades we also have corresponded near-daily on topics of global technology developments and their mission applications, particularly in US intelligence, commercial space sensors/architectures, and especially on all aspects of Maritime Domain Awareness (MDA). I served 26 years as a Naval and National Intelligence officer, nine years as a civilian program analyst and concept developer for the Naval Undersea Warfare Center (NUWC, Newport Division), and now I am an independent contract consultant for the Maritime Security Program in the Institute for Security Governance of the Defense Security Cooperation Agency.

I was the Director for Intelligence, Surveillance, and Reconnaissance (ISR) at the Navy Warfare Development Command (NWDC), and then the Intelligence Community liaison for the Chief of Naval Operations' Strategic Studies Group during 1999-2003 in Newport, Rhode Island—when Guy was Johns

Hopkins' Advanced Physics Laboratories Representative to NWDC, CNO SSG, and the US Naval War College in Newport.

Guy and I collaborated smoothly on ISR innovation projects for time-sensitive mobile targeting, missile defense, and MDA in five Fleet Battle Experiments for NWDC, influenced CNO SSG disruptive-futures ISR strategic concepts for Navy leadership, and he provided mentorship when I was the NWDC and CNO SSG liaison to the National Reconnaissance Office—Navy Coordination Group and to the National Security Space Architect. When I worked for NUWC, Guy continued to advise me on emerging technologies and over-the-horizon architectures to enable autonomous undersea, surface, and aerial systems to participate innovatively in Joint and Coalition wide-area ISR and targeting networks, including as distributive swarms. Guy's amazing technical and operational intelligence career across the US Air Force, the US Navy, US Coast Guard, and JHU-APL (Navy FFRC) has been a deep and wise resource for me and for the commands I served. His authoritative voice always reflects not just detailed all-aspect technical expertise, but also the warfighter relevance of one who conducted all levels of intelligence operations on the frontiers of the Cold War (and beyond) against America's adversaries in Europe, Asia, and the Middle East. In several cases, Guy hands-on designed essential US intelligence collection systems (e.g., in Rivet Joint, "Cobra-series," and other shipboard and airborne sensors), their performance requirements, their user-defined architectures, and also advocated for their acquisition and useful integration by military Services, Agencies, and allies.

What separates Guy Thomas as an extraordinary scientist and innovator is his understanding of "the Big Picture" and creativity in synthesizing technology with cogent requirements for a full spectrum of government and commercial users. He is also an action doer—a relentless implementer and able advocate for the merits of ideas and technologies—who does not take no as a final answer from even the seniors in the established order, and converts ideas from the laboratory into operational capabilities.

Our productive professional collaboration intensified immediately after 9/11/2001. We served together on a joint USN-USCG experts task force led by CNO N7 that assessed maritime vulnerabilities for Homeland Defense/Homeland Security—including likely threat scenarios and potential critical targets—and most importantly, scoped possible technical and operational solutions. The resulting ideas for national authorities were tested in near-weekly interagency wargames in Newport and Washington into 2002.

It was then that Guy crystallized his breakthrough innovative concept for a MARITIME NORAD. Thousands of airline, military, and general aviation aircraft active on any given day could be tracked, identified, navigated, deconflicted, and collisions avoided by means of Identification Friend or Foe (IFF) equipment, and by standardized international codes and procedures—and their tracks are linked to flight plans amplifying intent/purpose. This was critical to the Homeland Defense needs of the North American Aerospace Defense Command.

However, in 2002 there was no parallel networked international system to provide ubiquitous worldwide tracking, identification, and security knowledge for over 50,000 merchant ships above 1000 tons active daily on the world's seas, and for perhaps 350,000 or more smaller vessels of all categories. Guy assessed these needs around a gelling concept for MDA (aka Maritime Situational Awareness) for which he was a lead advocate. In 2003-2006, both Guy and I visited the USCG Research & Development Center at New London, CT, looking for emerging technologies of potential added value to the organizations we supported. Of special interest was a line-of-sight terrestrial technology, the Automated Information System (AIS), that offered many of the traffic-control tracking, identification, safety, and informational advantages of IFF for seaborne platforms, but was then limited to the horizon. Guy's innovation was to put the technology on satellites in space to make it globally usable, especially over the horizon.

He worked incessantly with industry and government to design the space payloads, the communications architecture, and to make the system affordable enough to be practically used by most of the world's shipping in most environments—all imperative requirements to transcend a technological innovation into a new paradigm and a new level of service and safety for all.

These efforts and innovation led to Guy rightly becoming the Chief Technology Officer for the USCG and a key technical advisor to the Department of Homeland Security, the Department of Transportation, and for the National Maritime Intelligence Integration Office in the 2000s. As the Naval Sea Systems Command senior technical liaison to NORAD and US Northern Command 2006-2011, I input much from Guy's concepts toward the May 2008 revision of NORAD's Charter, by which the Command first integrated the Maritime defense of North America into its primary mission, co-equal with Air and Space defense.

In the 2010s, Guy Thomas expanded his indefatigable energies into his second passion, mutually supportive to S-AIS—use of commercial space sensing for unclassified wide-area ocean monitoring and open-source data services. AIS is eminently useful as long as it is used by the vessel involved and it wants to be tracked.

However, transnational maritime criminal activities such as narcotics smuggling, human trafficking, and especially illegal fishing in territorial waters and Economic Exclusion Zones have become the primary maritime security threats for most of the world's countries, even landlocked African countries that depend on fisheries in great lakes of the Great Rift Valley. Criminals at sea do not want their vessels tracked.

To find and prosecute them, most countries lack resources for large fleets of ships and aircraft to patrol their waters, nor for launching their own expensive satellite sensor constellations. In creating C-SIGMA, Guy has been a major leading international advocate of increasingly capable (coverage, resolution, day/night/all-weather, revisit, processing, latency, data dissemination, display, and storage & retrieval) satellite-based monitoring with radar, infra-red, electromagnetic, multi-spectral, and hyper-spectral sensors that supplement S-AIS data to detect, track, characterize, and identify uncooperative "dark targets." Private

Sector competition and investments are making these technologies' on-demand user-tailored data services available to almost all nations desiring to gain MDA for the governance of their waters. This includes observing pollution and other anomalous maritime environmental conditions.

In an era when narco-traffic and human smuggling are increasing, 90% of the world's fisheries are depleted or under severe stress, and the oceans are deteriorating from pollution and climate change the MDA that commercial space services are providing are of tremendous value in developing the mitigating responses and solutions. Guy Thomas is a prime proselytizer of the worldwide revolution in space-based ocean monitoring, available to anyone with an Internet connection and a credit card.

I am now using frequent advice and vectors from Guy in developing strategic maritime security education courses for senior interagency officials of America's multinational friends, supporting the Institutional Capacity Building objectives of the Defense Security Cooperation Agency and the US Government. In providing our allies full-spectrum MDA education, I am orienting them on the availability and affordability of commercial space maritime data services, and how they can integrate these into their MDA operations, analysis, and knowledge. Over time, such MDA is making a significant contribution to their food security from illegal fishing and national security from drugs, contraband, and terrorism.

It has been a privilege to work with and learn from Guy Thomas, and to be present at his creation of the "Maritime NORAD," S-AIS, and C-SIGMA—from early concepts through to widely-used capabilities—for the immense benefit of US national security and for the unprecedented integration and safety of global maritime commerce.

—Julio J. Gutiérrez (Capt., US Navy (Ret.); Maritime Security Consultant)

Nomination by Keith Masback
(Former Chief Executive Officer of the United States Geospatial Intelligence Foundation)

Dear Committee Members,

I enthusiastically and unhesitatingly recommend George Guy Thomas for the National Medal of Technology and Innovation in recognition of his visionary and undeterred efforts to bring the space-based Automatic Identification System (S-AIS) into existence.

I write to the Nominating Committee as a 30-year national security veteran who has served as an officer in the United States Army and as a member of the Senior Executive Service at both the Army and at the National Geospatial-Intelligence Agency (NGA). During my government service tenure I was responsible, variously, for all future planning and programming for Army Intelligence, all future planning, requirements, and programming for NGA, and for the 24/7 /365 operational employment of our nation's classified imaging and missile warning satellites. For a decade after my government service, I led the international educational and trade association for geospatial intelligence technologies. I had the pleasure to work with Guy on and off for the latter 20 years of my career.

It is from this perspective that I share a glimpse into the enormity of Guy's technological innovation. S-AIS created the possibility of global Maritime Domain Awareness. The new transparency afforded by this technological advance means that every vessel is equipped with the ability to see and be seen, wherever it is operating. The unique maritime safety implications of this are obvious, as are applications related to the supply chain, especially with regard to the global petroleum trade.

Additionally, please note that thanks to Guy, our nation has been information-enabled to successfully preclude the proverbial 'rusty freighter' sailing into New York Harbor with a nuclear device. Thanks to Guy, all nations are empowered to interdict the illegal fishing operations which threaten our food chain. Thanks to Guy, maritime polluters can be identified, tracked, and held accountable for their infractions. Thanks to Guy, international sanctions can be reliably enforced because of the ability to monitor activity of vessels which are broadcasting on AIS and to rapidly seek out those that have 'gone dark.'

S-AIS is a fundamental technology that impacts every single American's life every day. It's akin to the Global Positioning System in that regard. It only exists because of Guy Thomas. I urge the Nominating Committee to appropriately recognize the quiet professionalism of Guy Thomas, who never toiled for glory or profit, but rather dedicated his tireless efforts dedicated to enhancing the safety and security of our nation and the globe.

—Keith Masback (Former Chief Executive Officer of the United States Geospatial Intelligence Foundation)

Nomination by Dirk Van de Ryse
(Director, Situational Awareness and Monitoring Division, European Border and Coast Guard Agency)

Dear Committee Members,

I recommend George Guy Thomas for the National Medal of Technology and Innovation because of his extraordinary vision and persistence in bringing space-based Satellite Automatic Identification System (S-AIS) to the world.

In my charter as Director of the Situational Awareness and Monitoring Division, I am responsible for the provision of reliable and comprehensive situational awareness. In this role, I oversee the mobilization and fusion of a wide range of sources of information, including geospatial data, and encourage the application of innovative approaches towards the collection of actionable intelligence, with a view of reducing the vulnerability of the external EU borders.

Frontex' mission is to help European Union Member States implement EU rules on external border controls and to coordinate cooperation between Member States in external Integrated Border Management. The security of the maritime borders of Europe are of primary interest to the EU and we follow research relevant for the control and surveillance of these borders very closely. It would be hard to overstate the impact satellite AIS (S-AIS) has had on our operations and it was with great delight I met the man responsible for its creation, Guy Thomas, several years ago.

His writing over the last 18 years on space-based maritime awareness and how to employ S-AIS in conjunction with other sensors and platforms, both space and terrestrially based, for the safety and security of the global maritime domain have been a primary guidepost to not just us, but the entire marine world.

S-AIS is now used to significantly improve maritime safety and security, as well as protect its environment and resources, globally but especially here in Europe. It is the foundation on which we have built our Maritime Domain Awareness system to counter smuggling of all types, be it people, narcotics, weapons or contraband. It is also used to identify ships which we detect pumping their bilges in illegal waters. Because many ship captains now know we can detect and identify their illegal dumping of fuel in European waters.

—Dirk Van de Ryse (Director, Situational Awareness and Monitoring Division, European Border and Coast Guard Agency)

C-SIGMA Conferences

Since April 2005 what is now C-SIGMA1 has held 11 highly successful international conferences at the following locales:

2005-6 USCG HQ., Washington, DC; (TEXAS 1-2)

2007-9 Canadian Embassy, Washington, DC; (TEXAS 3-4)

2010 ESA Earth Observation Centre, Frascati, Italy; (C-SIGMA I)

2011 NATO's Centre for Maritime Research and Evaluation, La Spezia, Italy; (C-SIGMA II)

2012 Canadian Embassy, Washington, DC (C-SIGMA III)

2013 Irish National Space Center, Cork, Ireland; (C-SIGMA IV)

2014 JAXA and the Japanese Coast Guard), Tokyo, Japan; (C-SIGMA V)

2015 Inmarsat HQ, (hosted by England's Space Catapult). (C-SIGMA VI)

2017 European Maritime Safety Agency, Lisbon Portugal. (C-SIGMA VII)

These conferences have been attended by all the significant Earth observation and AIS satellite builders and operators, and most if not all, of the builders of dynamic data analysis software, focused on Earth observation. Many users of this data from all corners of the globe have also participated.

The first four conferences were called TEXAS (Technical EXchange on AIS via Satellite.

Acknowledgments

I would be remiss if I did not acknowledge the sustaining support I received from several sectors.

First must be Greg Flessate, head of business development at ORBCOMM. He instantly saw the potential business potential of Satellite AIS in early October 2001, and his unwavering support and that of the ORBCOMM team he led over the next 10 years were crucial to the success of Satellite AIS. To this day, he is still hard at it. Jeff High, a US Coast Guard Senior Executive Service member, also immediately saw the global potential of Satellite AIS during my initial brief to him, in November 2003. He gave me both the funds and unlimited support in the face of political and procedural obstacles.

S-AIS would not have happened anywhere near as early as it did without the aid of Greg and Jeff. I briefed well over a thousand people between my first meetings with those two men. About 85 percent either could not see the utility or had no money. Many, maybe the majority, had neither money nor vision. It got very discouraging.

A short step behind them would be Pete Wilhelm, a former Navy Center for Space Technology director. His belief that we were on the right path and his guidance was crucial to me. He believed I was on the right course long before anyone else in the national space hierarchy did.

Finally, I must single out the excellent support I received from my former employer, Johns Hopkins University Applied Physic Laboratory, during the first two years of this journey. I could not have persevered for the first three years without the strong support of Russ Gingras, the man who hired me at the Applied Physics Laboratory in 1995. He and I had commenced working together in 1985. I was on the research staff of the Naval War College where I, with his substantial help, designed and led the first command and control countermeasures wargames held anywhere that addressed space system vulnerability in any detail. We had all the three-letter agencies plus the National Security Council looking over our shoulder for all four games, especially the last, held at the Applied

Physic Laboratory's Warfare Analysis Lab. Russ was the APL team leader for all four games, and you could not have asked for a better counselor and collaborator. His strong support and guidance during my last years at the Laboratory were the rock I clung to in those turbulent times. That superb support continued over the next two years (2003-2005) after I left Johns Hopkins as we finally got funds for S-AIS (just after I left). Jeff High's team was building the implementation plan of the United States for Maritime Domain Awareness. I was his assistant, and Russ was my mentor.

In May 2004, Russ and Jeff collaborated to hold the first and only Maritime Domain Awareness Summit of governmental agencies in the United States. It was for government officials only, and Jeff held it in the Warfare Analysis Lab. The room had seats for exactly 100 principals, each assigned by seniority. The 100th person in seniority was the director of Naval Intelligence, a two-star admiral.

My two immediate bosses at the Applied Physics Laboratory, Steve Biemer and Chris Latimore, were also significant sources of support and wisdom. Many others at the Applied Physics Laboratory were too, but there are too many to single out. It was a team effort. However, I must acknowledge the exceptional support I received from the graphics shop. Their wizardry with graphics art was instrumental in selling the ideas of S-AIS and Maritime Domain Awareness. Years later, I was longing for their skills as I worked to make the C-SIGMA concept understood.

Guy Thomas

notes

notes